RELIGION
OF
SPORTS

NAVIGATING THE TRIALS OF LIFE
THROUGH THE GAMES WE LOVE

Gotham Chopra
AND Joe Levin

ATRIA BOOKS

NEW YORK   LONDON   TORONTO   SYDNEY   NEW DELHI

ATRIA
BOOKS

An Imprint of Simon & Schuster, Inc.
1230 Avenue of the Americas
New York, NY 10020

First Atria Books hardcover edition December 2023

ATRIA B O O K S and colophon are trademarks of Simon & Schuster, Inc.

For information about special discounts for bulk purchases, please contact Simon & Schuster Special Sales at 1-866-506-1949 or business@simonandschuster.com.

The Simon & Schuster Speakers Bureau can bring authors to your live event. For more information, or to book an event, contact the Simon & Schuster Speakers Bureau at 1-866-248-3049 or visit our website at www.simonspeakers.com.

Interior design by Kyoko Watanabe

Manufactured in the United States of America

1 3 5 7 9 10 8 6 4 2

Library of Congress Cataloging-in-Publication Data

Names: Chopra, Gotham, 1975– author. | Levin, Joe, author.
Title: Religion of sports : navigating the trials of life
through the games we love / Gotham Chopra with Joe Levin.
Description: First Atria Books hardcover edition. | New York, N.Y. :
Atria Books, 2023. | Includes bibliographical references and index.
Identifiers: LCCN 2023001677 (print) | LCCN
2023001678 (ebook) | ISBN 9781501198090 (hardcover) | ISBN
9781501198106 (paperback) | ISBN 9781501198113 (ebook)
Subjects: LCSH: Sports—Religious aspects. | Athletes—Conduct
of life. | Sports spectators—Conduct of life.
Classification: LCC GV706.42 .C56 2023 (print) |
LCC GV706.42 (ebook) | DDC 248.8/8–dc23/eng/20230126
LC record available at https://lccn.loc.gov/2023001677
LC ebook record available at https://lccn.loc.gov/2023001678

ISBN 978-1-5011-9809-0
ISBN 978-1-5011-9811-3 (ebook)

To true believers everywhere . . .
See you in the stands.

Spirituality means a lot of different things to different people. For me it's your deepest purpose. I do want to know the whys of life. I want to know why we're here. Where we're going. Trying to find that deeper purpose. To live it through sports in a very authentic way makes so much sense to me. Having these dreams and goals and aspirations, and waking up and putting in the work, and miracles happening, and all this magic that sports creates, and I'm in the middle of it. I get to live that through sports.

—Tom Brady

Contents

Introduction

ONLY 1.6 SECONDS remain in Game Seven of the NBA Finals. Your team is about to inbound the ball. You're down one. A basket wins the championship.

It's the Super Bowl. Your team is three yards from the goal line with just enough time to run a final play. You're down by four. It's all or nothing.

We're in extra time at the World Cup. Your team is down a goal, lining up for a corner kick. Everything relies on the foot of a nineteen-year-old striker. The whole world holds their breath.

Your stomach is inside your throat. Your breathing is staggered. You can feel every heartbeat as it races and vibrates through every cell of your body. You look around to see if anyone else feels this way. They do. It's euphoric and agonizing all at the same time.

You want to cry.

You want to scream.

You want to pray. You invoke God. You make a promise and commit to being better—even though you don't believe in Him.

////////

You're in the stadium with 43,607 other fans who feel just like you. They're Black. They're White. They're Asian and Aboriginal and Arab and Israeli and Christian and Jewish and Muslim, Buddhist, Hindu, atheist, gay, and straight. They are saints and sinners, virtuous and vengeful. They account for every type of human experience. Some have been fans since before you were born. Some are barely old enough to remember this moment.

Outside the stadium, millions more tune in by TV, radio, and live stream. They follow the action in crowded metropolises and tiny rural villages. They're just like you, hanging on to each second, having surrendered themselves over to something beyond their control and bigger than themselves.

And right now, you are all one.

The clock starts. The ball snaps. The striker kicks.

Time stands still. A singular moment encoded with infinite potential. The grand mystery of human existence captured in a flicker and flare of reality.

Hopes and dreams. Redemption. Deliverance. Heaven and Hell. Curses and miracles. They all await on just the other side of the moment.

It is the moment of truth.

It is revelation.

Sports *is* religion.

Believe.

///////

I was eleven years old the last time I peed my bed. If I'm being honest, the absolute last thing I would have thought when I woke up that night is that several decades later, this would be the opening scene of a book I would write. But here we are, and somewhere, eleven-year-old Gotham is even more embarrassed than he was back then.

I remember the exact date of the bedwetting, October 25, 1986, not because the incident was so traumatic but because all the events that led up to that indignity were. I was in Tenafly, New Jersey, visiting cousins who were avid New York Mets fans. A few hours earlier, my beloved Boston Red Sox played their Mets in Game Six of the World Series. I was positive that the night was going to be one of the best of my life; not only would my Sox *finally* be World Series champions, but I would get to taunt my cousins for the rest of the trip as well.

If you know anything about baseball history, you know that that's not how it worked out.

In the tenth inning, a weak ground ball dribbled between Sox first baseman Bill Buckner's legs, the game-winning run scored for New York, and I sat frozen in my cousins' living room staring stunned at the television, utterly demoralized. The "Curse of the Bambino" lived on.

I cried myself to sleep that night, I was so aggrieved. I couldn't bother to move when I had to use the toilet in the middle of the night. My mother, who had immigrated from India, didn't understand why I cared so much about all of the Boston teams. But God bless her, she consoled me back to bed and put the sheets in the washing machine before my cousins had one more piece of ammunition with which to make fun of me. When the Sox lost the series in Game Seven two days later, I told Mom I was convinced that the Red Sox—and I—would never know what it was like to win a World Series.

"If that's true," she told me, "then are you still going to be a Sox fan?"

I looked at her, puzzled. "Why wouldn't I be?"

////////

In the years that followed, I began (and have continued) to make annual pilgrimages to Fenway Park, and from each March to October, a

Sox cap became a part of my regular uniform. I suppose, looking back now, that I thought that through these little moments of allegiance, I could personally *will* the Sox out of misery, that my devotion could be the thing that finally led the Sox to the Promised Land. Or maybe I was just afraid of *not* doing those things, afraid that if I didn't wear my lucky cap or sit in my lucky seats, and the Sox lost, all of Red Sox Nation would turn toward me and say, "What on earth were you thinking? You blew it!"

How could I live with that type of shame? I couldn't. And so I repeated the same rituals and the same traditions year after year, including the most important one of all: I hoped.

Even when I moved to New York for college, I kept up the act. Of course I was repulsed by the sacrilegious locals and their affections for the Knicks and the Yankees, but I also knew that it was probably best if those same people *didn't* know that I was rooting for their archrivals. So I protested in smaller, more subtle ways than screaming, "The Yankees suck!!" in the middle of the subway. Instead I kept my head down and ordered Samuel Adams Boston Lagers on draft, feeling connected to home even while deep in enemy territory. Silent pride would have to do.

In 2003, years after I'd graduated from college and moved to Los Angeles, I thought it was finally going to happen—the Sox would win it all. We had a stacked team, with star sluggers such as Manny Ramirez and David Ortiz, aka "Big Papi," and powerhouse pitchers Pedro Martinez and Curt Schilling. One popular shirt at Fenway during that time honored our center fielder, saying, "WWJDD?" (What would Johnny Damon do?) I watched every inning of every game of the American League Championship Series at the famed Santa Monica watering hole Sonny McLean's, where Massholes congregated and cheered on their teams. So it was there during Game Seven that I watched "Aaron Fucking Boone" and the Yankees rip our hearts out once again (I'm sparing details because . . . well, I'm hoping you can understand why). I was in my late twenties by then, so I didn't wet the bed. But that postgame sensation of shame and sadness felt familiar to me, as did the acceptance that we would just have to try again next year. After the season ended, the team produced a slick highlight video. At the end of it, the screen faded to black, and a message appeared. "Still, We Believe," it read.

"We took the wording straight out of the Catholic canon," club president Larry Lucchino told *Sports Illustrated* of the slogan. "It's not 'We Still Believe.'"

Still, I *did* believe, even the next year, in 2004, when the Sox found themselves down three games to zero against the Yankees in the ALCS. No other team in baseball history had ever come back from a 3–0 deficit to win a series. Then, in Game Four, the Sox's Dave Roberts stole second, and Ortiz hit a walk-off home run a few innings later. In Game Five, Ortiz hit another walk-off, this time a line drive in the fourteenth inning. In Game Six, Schilling pitched seven innings with a bloody ankle.

Game Seven was as big a game as I can remember, in any sport. On ESPN's *SportsCenter* that night, Dan Patrick told New York superfan and acclaimed director Spike Lee, "When you're putting a movie together . . . you can try to write drama, but you can't write this kind of drama."

Spike interrupted Dan. "That's why movies, that stuff is fake," he said. "That's why sports is the greatest. Because it can't be scripted."

In Tokyo the day before that game, a woman visited an ancient temple when she saw a wall of prayer blocks. These blocks are small, wooden plaques hung at a shrine where, hopefully, some higher powers will acknowledge them. The woman had a block of her own to put on the wall, but before she did, she read some of the others. There were wishes for health and peace. For newlyweds and college graduates. For the suffering and forgotten. Then there was one that read, "May the Red Sox play always at Fenway Park, and may they win the World Series in my lifetime."

I cannot think of a better prayer than that.

The Red Sox, of course, did win. In that 2004 postseason, they won Game Seven to beat the Yanks, and then they beat the St. Louis Cardinals to win the World Series. After the final out, friends called me from all over New England so we could revel together, and I could hear church bells clanging in the background of their receivers, the celebrations ringing in every corner of New England. I cried when they—when *we*—won, and for weeks afterward, found myself muttering, "Can you believe it?" to nobody in particular. To this day, I can barely believe it was real.

Viewed one way, the Sox's victory is a classic sports story, an underdog tale for the ages. But viewed in another, it's a parable of faith—not just blind faith but rather the type that bonds communities, gives lives meaning, and connects people to something larger than themselves. The Sox World Series, when considered in this light, was not just a baseball game. It was a miracle on par with Moses parting the Red Sea.

Why? Because we Sox fans don't just root for a baseball team. We convene in something deeper than that, something more visceral and meaningful, something that causes you to say little prayers half a world away, from New York sports bars all the way to Japanese temples.

We are the true believers.

Believers in the Religion of Sports.

//////

One Sunday in 2020, the New England Patriots kicked a field goal early in the first quarter of a regular season game to go ahead 3–0. It wasn't a particularly important game, nor was it any sort of crazy kick. Everything about it was routine. Watching at home in Los Angeles with my son, whom I've trained to be as fanatical about the Patriots as I am, we hardly even acknowledged the three points. Maybe we clapped once or twice, but that was it.

My phone buzzed. It was a text from my mother. "2 a.m. here in Delhi," it read. "Go Pats."

My mom, Rita, grew up in India. When she moved to Boston and was raising me, she couldn't understand why the Patriots, who barely won a game at the time, had such a hold on me and my imagination. But here she was, years later, watching in the middle of the night from across the world, a family connected via football.

That's the power of sports.

While they often get dismissed as leisure or escapism, it's not fair to categorize sports as simple distraction. They're much, much more than that. Sports connect people, from entire countries to my family watching from opposite sides of the couch and the globe. Sports inspire, taking on meanings far beyond the scoreboard. Sports give us a place where we can see dreams come true. They help us heal. They show us how to get the absolute most out of our talents.

Put another way: sports are humanity.

Perhaps it should not be so surprising to me that my mother eventually grasped the power of sports for herself. Spirituality has been one of the constants in my family since long before I was born. I was raised on the frontier of an emerging New Age of spirituality that my father, Deepak Chopra, and his contemporaries forged. As a kid, I meditated. I watched my father speak at conferences. Before we went to bed every night, he told my sister and me, "Who are you? What is your purpose? Do you believe in God? And if so, how are you here to serve *her?*"

He always said "her." Every night, that's what I would think about as I went to sleep, and over breakfast, my dad would ask us, "Did you come up with any answers?"

For the longest time, I never did. My father talked about peak performance, being in the zone, and finding community in thousands of other souls all wishing for the same thing. He experienced those axioms through his faith. I looked for them there but never quite found them, at least not as viscerally as he did. But when I went to Fenway or watched Patriots games with my family, I felt all the same things. Over time, as I stewed over that question at night, I started to realize that through sports, I could speak the same holy language as my father—a different dialect, maybe, but the ideas remained the same.

Instead of the Cathédrale Notre-Dame de Paris, we have Notre Dame Stadium in South Bend, Indiana. Our holy war is FC Barcelona vs. Real Madrid, better known as El Clásico. We make pilgrimages to events like the Daytona 500. A team wins the Final Four and cuts down the net, creating a relic that will be cherished forever. Curses, like the Cubs and the Billy Goat, are created and lifted. Every seventh inning of a baseball game, we sing a hymn.

I could go on and on and on.

Religion, in the classical sense, is in decline. A recent Gallup poll found that for the first time in history, fewer than 50 percent of Americans are members of a church. Americans' belief in Christianity is down 12 percent in the last decade, and the share of the population that report themselves as being agnostic has risen from 17 percent in 2009 to 26 percent in 2019. In the last thirty years, the number of Americans who say they are not associated with a traditional, established religion has tripled.

But are we all just abandoning belief—full stop? I don't think so.

Not believing in *anything* would go against our nature. Consider this: scientists have proposed a so-called God gene, a part of the genetic code found in 90 percent of the population that prompts humans to believe in and seek out spiritual and mystical experiences. In other words, we are all predisposed to believe in something larger than ourselves. Gravitation toward tribes, rituals, and faith is quite literally baked into our DNA.

Here we have a contradiction. On one hand, Americans don't believe or participate in classical religion as much as they used to. On the other, humans *have to* believe in something.

It begs the question: What is everybody believing in?

By this point, just a few pages in, I'm guessing that you can already see where I'm going with this.

Sports are our new religion.

/////////

I've collaborated with many of the greatest athletes in the world: Tom Brady, Kobe Bryant, Serena Williams, LeBron James, and Simone Biles, to name just a few. While I'm working with them, I consider myself a "miner" more than anything, chipping away at their careers, myths, and greatest moments for nuggets of wisdom that can help us all lead better lives.

People always ask me what it's like to be in the presence of a GOAT like LeBron or Simone. And as much as my projects—like Kobe Bryant's *Muse* or *Man in the Arena: Tom Brady*—do indeed glorify their accomplishments, it's important to me that I cast light on the underlying vulnerability of the real people behind these extraordinary athletes. These GOATs are *not* perfect. Tom Brady doesn't eat kale *all* the time (but he does eat it a lot more than you and me). They all make mistakes, have regrets, feel darkness, and face challenges. What makes them extraordinary is how they navigate those trials of life, how they channel them into greatness on the field, the court, the pitch, and the mat.

So when I say that I consider myself a miner more than anything, those are the gems for which I'm always on the lookout. How does soccer player Alex Morgan handle the pressure of leading a team that is not only expected to be the best on the pitch but also act as trailblazers off of it? Why did surfer Kelly Slater's *fifth* world championship—and not the third, fourth, or even tenth—leave him feeling the most fulfilled? What made a bronze medal the most meaningful of skier Lindsey Vonn's decorated Olympic career? And how can I—a pretty average Indian-American guy whose own version of the NBA Finals is Sunday morning pickup basketball—use those lessons in my own day-to-day life?

My father taught me that a true spiritual practice is not just a relentless pursuit of an outcome but rather an observance and appreciation of process. So really, the truths that we'll discuss in this book are practical, prescriptive lessons that can be applied to our daily lives, resulting in a richer, more dynamic human journey. I am, after all, a miner: what follows are some of the prizes that I've uncovered over the years, nine chapters focused on nine truths that I've learned from a life lived at the nexus of sports and spirituality. They are nine principles that prove that sports are religion.

Like I said, sports have everything a traditional religion does, but they also have something that so many other religions do not. My father also taught me that any religion is just a set of rules and codes and stories that have little meaning—unless they are activated. And no faith is more active than the Religion of Sports, which packages all this wisdom in a way that anybody can understand it: wins, losses, championships, heroes, and underdogs. The Religion of Sports gets its power from being something we can all inherently *feel*, something with which we all can participate—something with which we all *do* participate. Sports is the religion where our gods are flawed humans like the rest of us; where anybody, with the right amount of luck and skill, can become a champion; It's a faith that creates miracles all the time. Sports help us look toward the infinite and believe. In ways both small and large, they help us unlock the riddles of life.

I mean, in every other faith, you have to wait an entire lifetime for salvation. Sports fans experience it with every win, every championship. Or, for Sox fans like me, it takes eighty-six years.

But trust me—it's worth the wait.

CHAPTER 1

Baptism

*All spiritual journeys begin with a moment
of divine inspiration . . .*

People are always coming up to me. "Jim, can you remember that goal against West Brom in 1968?" and I say, "No." But that's all right because they only want to tell you about what happened to them, anyway. "Well, you had the ball on the halfway line, and I remember that because I was with Charlie, and we'd just got two pies . . ." and it turns out the real story is about Charlie dropping his pie and what you did wasn't all that important anyway. I prefer that, really.

—Jimmy Greaves,
Tottenham Hotspur star from the 1960s

IF I'M WRITING a book about the Religion of Sports, there's only one real way for me to start it, right?

Clears throat . . .

Cracks knuckles . . .

Tries to imitate that pliability routine Tom Brady taught me . . .

IN THE BEGINNING . . .

///////

Everybody has their own story of initiation into any kind of faith, a moment where the whole world comes into focus just a little bit more, where a tribe welcomes them with open arms, and a new journey begins. A Jew's bar mitzvah. A Christian's baptism. A Hindu's *upanayana*. Many sports fans are born into their faith; in the same way that Muslim parents whisper a prayer into their newborns' ears, diehard Michigan fans will take their daughter home from the hospital directly to a room painted maize and blue.

I didn't have that.

My parents weren't sports fans. They didn't have the slightest clue why Americans lived and died with every victory and defeat. Mom and Dad (or Papa, as my sister and I called him) emigrated from India in 1970, the same year they married. My father, Deepak, practiced medicine, which made sense: it was the family trade. His father was a cardiologist and a really good one at that, becoming a lieutenant in the army and the medical adviser to Lord Mountbatten, the British viceroy of India before independence.

But Papa didn't always want to be a doctor. When he was young, he dreamed of being a storyteller, inspired by his mom, who would stay up late at night with him recounting Indian myths and legends. He loved reading, and once as a teenager, he picked up a Sinclair Lewis story where the hero was a doctor. Dad told me later that he felt like the character in the story was a god, just like the figures from the myths he listened to late at night. He said that he couldn't believe that everyday people could do such extraordinary things. It was a feeling I would recognize on my own soon enough.

Papa graduated from medical school in Delhi and immediately went to work in the field. The higher-ups assigned him to a small, rural village. Every time it rained, the lights flickered and then shut off. He loved the work and loved being able to help people. But my father is a learner, and he wanted to continue his education by studying Western medicine. So my parents traveled to Sri Lanka, Papa passed his exams there, and the two of them arrived in America shortly thereafter.

It was the height of the Vietnam War, and American hospitals faced a dearth of doctors. They welcomed foreigners with open arms, and that's how Papa found his way to Plainfield, New Jersey, for an internship. If their journey ended there, I may have wound up a Mets fan like my cousins (God forbid), but luckily, a year later in 1971, my dad started his residency at the Lahey Clinic in Boston. I came along a few years after, and I'm sure everybody in my family assumed that, one day, I would grow up and become a doctor, too.

That, obviously, is not what happened.

Growing up, my memories of my father are that he was always working, earning, and pushing the envelope so that my sister and I could go to private schools and eventually great universities. As it had been for him and his brother growing up, education would be the doorway to a future of unlimited possibilities for us. But as a consequence, I never played catch with my father. We never shot hoops or went to the batting range or drove balls at the local golf club. He didn't take me to games, and not once while growing up did we sit down and watch the Super Bowl or World Series or a knockout game in the NBA Finals.

I would have to find my faith on my own.

///////

Just as converts to a new religion have their initiation, every child of immigrant parents has their own story of assimilation. *How do you feel like you belong?* Very quickly, I found my answer in sports. Sports were like a language to me. They were how I learned how to be an American.

As far back as I can recall, I remember traveling around the Boston area and seeing Celtics, Bruins, and Red Sox gear everywhere (the struggling Patriots were sadly a source of shame most of my childhood, but we'll get to that in a later chapter). I loved all the Boston teams—I still do—but my first love and biggest passion was the Celtics. Can you blame me? These were the days of Larry Bird, Robert Parish, and Kevin McHale. The Celtics were the biggest act in town, one of the best teams of all time. How could a kid not fall in love with that? We would drive to the mall, and I would see huge billboards of Larry Bird selling Converse and think that if I could just get a pair of those sneakers, maybe I would be like everybody else.

At home, I commandeered the TV remote away from the rest of the family to put on the Celtics any chance I could. One time, Bird hit a three, and as I cheered on the couch, the camera cut to the crowd at the old Boston Garden. Papa looked at the screen, puzzled. On the TV, a group of fans screamed, all clad in Larry Bird jerseys.

"Why do their jerseys say 'Bird'?" Papa asked me. "They are not Larry Bird. Are they pretending to be him?" For him, the idea of a grown man wearing a jersey with another man's name printed on it was puzzling, if not a little absurd.

I was so relieved that none of my friends had come over to watch with me. "No, Papa," I said. "They just love him."

Another time, when I was about six, the Celtics played the Philadelphia 76ers. Back then, Celtics fans' disdain for the Sixers was second only to their scorn for Magic Johnson and the Los Angeles Lakers. The Sixers had Julius Erving, Moses Malone, and a rookie named Charles Barkley. If you've ever heard Barkley complain about how players in the NBA don't compete against each other like they used to, you can bet that he's probably remembering those old games against the Celtics. These teams *hated* each other, and as the two top teams in the East, they always seemed to be on a collision course in the playoffs. Games got chippy. Like, really chippy. Case in point: Once, during an *exhibition* game, play had to be stopped three different times because of fighting. The Celtics legendary general manager Red Auerbach ran onto the court and up to Malone. "Hit me!" he said. "Go ahead, hit me. I'm not big."

Luckily for Auerbach and the future of the Celtics dynasty, Malone showed a little restraint that night.

So, when the Sixers visited the Garden that night early in the 1984 season, tensions already ran a little high. I wore my Celtics shirt to school earlier that day (much to Papa's chagrin) and rushed through dinner, hoping that I could make the game start just a little bit quicker if my green beans disappeared in record time. Finally, tip-off came. I watched as Bird made basket after basket, while grabbing rebound after rebound. Through thirty minutes, he amassed 42 points along with 7 rebounds. He locked down Dr. J on the other end of the floor. Erving had only scored 6 points all game.

Then things went totally off the rails.

Bird and Dr. J got tangled up, and Bird got called for an offensive foul. Bird screamed at the refs. I yelled at the TV at home. Every single person in the Boston Garden booed. Next thing I knew, it was pandemonium. Malone tried to hold back Bird, and it looked like Malone was choking the Celtics star. Pretty soon, Dr. J and Bird exchanged punches, two of the best players in the world in a fistfight during the middle of the game. I couldn't believe it. It was the Titans fighting the Greek gods for Mount Olympus. Who would win?

And perhaps most confusing as a six-year-old: Who should I root for? I *loved* Larry Bird, wanted his shoes, wanted to be just like him. But Dr. J looked more like me. He was the one with darker skin like mine, not Larry Bird. I wondered, was my allegiance in the wrong place?

I shoved those thoughts aside pretty quickly. I was a Boston fan.

That was my community. It was a community where I was welcome, too. Larry Bird was my guy—was our guy—and it didn't matter whether he was white or blue or green. Nor did it matter what color I was, either. We all wanted the same thing.

We were all part of the same tribe.

////////

My Celtics fandom only grew from there. I watched every single game I could. I indoctrinated myself into the lore. I learned that the Garden got its signature parquet court pattern because during World War II, when the country faced a wood shortage, the Celtics could only get pieces of red oak broken into a medley of different shapes and pieces. They cobbled together whatever they could. After the war ended, they decided it actually looked all right, and the resulting crisscross pattern is a tradition that's continued ever since. I learned that the team's name came from the generations of Irish immigrants in Boston and felt even more connected to the team because of it. If the Irish could get their own team, maybe one day we could have one, too. I checked out books from the library about the great Celtics teams of yesteryear and argued with friends about the greatest teams, acting as if I knew what I was talking about. "McHale is great! But he's definitely no Bill Russell. Nobody could rebound like number six." Of course, Russell retired years before I was born.

The Celtics became tied to everything I did—even during after-school playtime.

I'm thinking now of the classic scene in the film *One Flew Over the Cuckoo's Nest* where Jack Nicholson's character, R. P. McMurphy, insists on watching the World Series while cooped up in an Oregon psychiatric hospital. Nurse Ratched, played by Louise Fletcher, refuses to bend the rules and turn on the TV, even after McMurphy gets other patients to vote to watch the game with him. McMurphy, steaming mad, sits down on a bench, staring at his own reflection on the turned-off screen, when he launches into what has to be one of the greatest play-by-play calls of all-time. He closes his eyes and creates a world of sports out of thin air.

> Koufax kicks . . . he delivers. It's up the middle, it's a base hit! Richardson's rounding first! He's going for second! The ball's in to deep right center! Davidson, over in the corner, cuts the ball off! Here comes the throw. Richardson's around the dirt! He slides, he's in there. He's safe! It's a double! He's in there, Martini!
>
> Look at Richardson, he's on second base. Koufax is in big fucking trouble! Big trouble, baby! All right, here's Tresh. He's the next batter! Tresh looks in. Koufax . . . Koufax gets the sign from Roseboro! He kicks once, he pumps . . . It's a strike! Koufax's curveball is snapping off like a fucking firecracker. Here he comes with the next pitch. Tresh swings! It's a long fly ball to deep left center! It's going! It's gone! . . .

The patients go nuts, cheering like they'd just won the World Series themselves, and so did I the first time I watched the movie. What I love about that scene is that it captures the part of being a sports fan that is beautifully delusional. Even when we can't watch sports, we're drawn to them, wanting to create that sense of familiarity and routine, that feeling that anything can happen—that even the great Sandy Koufax can give up a home run.

That's a fictional example, but others find peace while visualizing sports as well. In 2011, Japanese astronaut Satoshi Furukawa was part of a mission aboard the International Space Station. In an interview before he was launched to space, he was asked what it was like to finally complete his training. "I used to be a member of a baseball club," Furukawa said, "and it's like in prep for a game I have been practicing and training [for], like swinging bats every day. . . . So finally it's my chance to go."

One night, floating above earth, cameras found Furukawa alone in a corner of the space station. He had a baseball in his hand and pitched it down a corridor. Then he pushed himself off the wall, and flying through the air in zero gravity, raced to the other side and picked up a baseball bat before the ball reached there. He took a swing and then was racing back to play the field. Furukawa was playing baseball all by himself, his way of reminding himself of home.

For others, it's the ballpark that they miss. Barry Rosen worked as an American diplomat in Iran in the late 1970s. When a group of students overran the embassy in 1979, Rosen became one of the fifty-two Americans taken during the Iranian hostage crisis.

Rosen came from Brooklyn, New York. As a kid, he made pilgrimages to Ebbets Field with his father to cheer on Duke Snider and the rest of the Dodgers from the right field bleachers. His mom would pack them a lunch, and the father and son would sit there watching the game, chatting with the people around them, enjoying the afternoon.

When he was taken hostage, the captors beat and threatened Rosen. At one point, they moved him to solitary confinement. They didn't allow him outside for nearly four months. But there was one part of the day that Rosen always looked forward to: Once a day, he would close his eyes and conjure up Ebbets Field from memory. Some days, his seats would change, and the characters around him would, too. Other times, his father would sit next to him, and they'd unwrap their lunches and chat about the pennant race. He imagined every pitch, every swing of the bat. He would see Dodgers outfielder Carl Furillo fielding line drives off the wall and miraculously firing out the runner at home base. Rosen did this every day, playing entire seasons in his head from his jail cell.

"Baseball," he told the *Los Angeles Times* decades later, "kept me sane."

At least, the *idea* of it did.

I didn't need basketball to keep me sane. The stakes were much, much lower for me than they were for Rosen. But I did need sports to keep me busy, and so, nearly every day after school, I would come home and fire shot after shot at the basketball hoop in the front yard. To me, I wasn't in my driveway. I was at the old Boston Garden shooting underneath the banners. I wasn't a kid trying to find my place, trying to figure out how I could be American like my friends, Indian like my family, and whether I could be both at once. I was a *Celtic*. I wrote out every roster in the NBA, and my cousin Bharat and I spent days playing a full season's worth of made-up games. We would mimic the way the best players in the league shot—turnarounds, free throws, and baby hook shots. We'd try (and fail) to dunk. We'd talk to ourselves the entire time we performed the charade, impersonate the legendary Johnny Most, the Celtics' ornery play-by-play announcer, envision ourselves in the sold-out Garden, and mimic both the referees and the brutish fans that castigated them—all at the same time.

I'd dribble the ball at the edge of the driveway, talking out of my throat to impersonate the raspy voice of Most. "It's all coming down to this," I—or Most—would say. "Tie game and DJ [Dennis Johnson] is bringing it up. This is the Celtics' last chance. . . . " Bharat would roar like the Garden crowd, and as he moved to set an imaginary pick he would scream, "Let's go Celtics!" over and over, out of the corner of his mouth. I came off the pick, dished the ball to Bharat. "McHale has it now, and what's this?" He passed me the ball as I cut toward the hoop. "It's a give-and-go! DJ is wide open for the layup, and it's . . .

"*GOOD!*"

Bharat and I hugged under the basket. He gave me a postgame interview. Then we consulted our schedule, figured out what game was next, and started it all over again. Miraculously, the game always came down to the final shot when the Celtics had the ball. And if that shot didn't go in (Danny Ainge was a notoriously streaky shooter), then the referees would huddle and realize that they made a mistake and actually ... *Wait! What's this? Upon further review, there's one second left! The crowd is going crazy. The Celtics have another chance to win the game! Bird for the winner.*

The Celtics *always* won the game.

This type of daydreaming is a central trait for any sports fanatic. I see it now in my son whenever he and his friends fire up the PlayStation and create a Madden franchise or run through a full tournament in FIFA. It's like our sports fan superpower: When all hope is lost in real life, we simply dream up a new way out. When we're down two goals in stoppage time, we can always imagine that a miracle could happen, that one goal could follow another until that moment of pure euphoria erupts. It's one of my favorite parts of watching sports broadcasts on television—seeing how invested fans get in a close game, clasping their hands, dawning rally caps, making deals with some higher power to just help them out on this one game. Who among us hasn't done this? Why? Because when we close our eyes, we are transported to this new type of place, some mix of hope and meditation.

It's our way of saying a prayer.

/////////

My guide, my shaman if you will, into the mysticism of the Celtics was a man named Alan Rosenfeld. Alan was an entrepreneur, and he owned his own auto glass business and had become friends with my parents. At some point, my dad confided in Alan that I'd become an obsessive Celtics fan. He noted that he didn't quite understand what was so special about these games or why so much idolatry formed around the players.

"You know," Alan told him, "my family and I have season tickets. Let me take him to a game."

When Papa came home and told me what Alan said, I nearly froze. *Me? Going to Boston Garden? No. Way.*

I scrambled up to my room to check the Celtics schedule. *Do you think we're going to go to the game against the Knicks? The Bulls? Maybe even . . . the Pistons?* (Imagining going to an actual Lakers game was just too much to even fathom.) I tore through my dresser, trying to figure out which pair of green shorts should go with which green shirt. *Hat or no hat? Should I make a sign?* I stood in front of the mirror and practiced my cheering.

A few weeks later, Papa came home and announced, "Mr. Rosenfeld asked if you can go to the game next week." And for the next week, I barely slept a night.

The day came, and at school, I became a maniac. In class, I drew scenes of Larry Bird hitting a three, with me watching on from the stands next to Alan. During recess, I bragged to everybody that in just a few hours, I would come face-to-face with the Celtics. I wondered if it could possibly be as special as I was imagining—but of course it could. This was the Garden. These were the Celtics. For a kid, it was like spending a lifetime reading Batman comics and then being offered a tour of the Bat Cave. How could anything be better?

Alan picked me up, and he quizzed me the whole way there: Who was my favorite player? How did I like our chances this year? What did I think about this young kid on the Bulls, Michael Jordan? I said that he seemed like he would be pretty good one day but that the Celtics surely didn't need to worry about him. We had Larry Legend, after all.

Then it was my turn to quiz Alan. Where were our seats? What type of concessions should I get? How loud was I allowed to cheer? I asked him how many games he had been to, and when he told me that he had been to so many that he lost count, that he'd been to almost every Celtics game of the last decade, that he was *at* the game where Larry Bird and Dr. J traded blows, it stunned me so much that I couldn't even stammer a response. One day, I swore to myself, I would be just like him.

And then we were there. It was real. The Boston Garden.

Workers broke ground on the Garden in the late 1920s, the whole project the brainchild of a man named Tex Rickard, who just might be as important as anybody in establishing the American sports system that we know today—although you've likely never heard of him. Rickard was a boxing promoter first and foremost, the first man to organize a prizefight that featured a million-dollar purse, and he was the type of colorful character that can only exist in sports. A native Texan, one of his first jobs was to be the marshal of a tiny town in rural Texas. Then he got gold fever, joined the Alaska gold rush, and though he didn't strike pay dirt, he did find a more lucrative career owning and operating a saloon in the Klondike. Once the gold craze shifted to Nevada, Rickard picked up and moved his operation there. Slowly, he learned the power of effective promotion. Rickard soon took his talents to New York, drumming up interest in boxers, and his business grew and grew. Eventually he needed somewhere to stage his biggest fights, and so he helped organize the construction of the third iteration of Madison Square Garden—the one that is still considered a mecca to this day. That arena was so successful that Rickard convinced several Boston businessmen to build a "Boston Madison Square Garden" as well. Eventually the name would be shortened, and the Boston Garden would be home to the Celtics, Bruins, and all of the greatest spectacles to come to the city. In 1937, workers erected a ski jump ramp inside the arena, selling tickets promising that "Norway has come to Boston!" The very first non-sports event in the brand-new arena in 1929 was a conclave for a traveling British evangelist. When Rickard died later that year, his funeral was attended by more than ten thousand people—and although he was never religious during his life, several tribes claimed him as their own. His final rites were read by, according to *Sports Illustrated*, "a Baptist clergyman, an Episcopal clergyman, a Catholic lawyer, and a Jewish judge." Maybe they should have just found a referee to do the honors.

Rickard's Boston creation didn't look like the sleek modern arenas that are so pervasive around the world today. It sat in the middle of West End, an otherwise nondescript neighborhood. Cabs swerved outside. A train pulled up on an elevated track directly across from the entrance. Some say this huge building rose like a barn or a factory, but I always thought that it looked like some kind of church. Something like spires rose on either side of the building, connected by a series of huge windows. They could have been stained glass for all I cared. In the middle of it all when I visited that first day, on top of the building, was a giant Budweiser billboard, the Clydesdales running in formation.

Every religion has holy places—temples, shrines, cathedrals. In fact, it seems like building monuments to the sacred is deeply rooted in human nature; in Turkey, archaeologists found ruins of a temple dating back to 10,000 BC. In the Middle Ages, cathedrals might have been built on the outskirts of town, but quickly, buildings would start popping up around them. Often, a city's center would actually move to surround the church. And if the city had grown so much around it that a larger cathedral was needed, they'd knock down the old one and rebuild something bigger and better on the same holy ground. It's not unlike the way that downtowns can be revitalized by a new stadium—think Cleveland and Rocket Mortgage FieldHouse—or how the new Yankee Stadium was built right next to the old one.

As we walked closer to Boston's basketball cathedral that day, I tried to take mental snapshots of everything, Alan guiding me to make sure that I didn't knock into anything or anyone in my overwhelmed daze. I wasn't watching where I was going; my eyes were studying the world being revealed all around me. I had seen the Garden so many times on my television. I couldn't believe any of it was real. Alan broke the trance, and tapped my shoulder. "Smell that?" he said. Something salty, buttery, and delicious wafted toward me. He pointed toward a red and white shack. "That's Buzzy's Roast Beef. We'll go after the game."

Ah, yes. Communion—that sacred food and drink that you consume to symbolize your faith. For some religions, it's wine. For others—well, roast beef can prove your devotion just as well.

Everyone, everywhere dressed just like me: green on green on green, shamrocks emblazoned on every piece of clothing imaginable. These were my people. We handed over our tickets, and then we were inside the dark, cramped corridors. Alan led me through the maze, and I tried to memorize the steps to the seats. Soon the route would become second nature—or at least I hoped it would. We worked our way past the concessions until we got to our section, and then there it was: the court, the banners, the Celtics—the actual Celtics!—down there warming up their jumpers.

"Well, what do you think?" Alan asked me.

It was overwhelming—not in a bad way, but in the way where things feel so special, so surreal, that you're convinced they can't be true. It was like a jolt running through me, energizing me. I felt it in my stomach. This was somewhere I belonged.

"DJ," I said. "He's . . . right there." Robert Parish. Kevin McHale. Danny Ainge, Greg Kite, Scott Wedman, Rick Carlisle, Cedric Maxwell, M. L. Carr. And wait a minute, cue the Gregorian chants: Larry Freaking Bird—not even a hundred feet away.

He was about to get a whole lot closer than that.

//////

Alan kept an eye out for me. A few times a year, he drove up to our house, and I'd bound into the backseat. The whole way to the game, we would talk about the team and the league. We'd talk about other Boston teams, too, and in those days, it looked like the Sox might actually finally do it. He asked me if I was playing sports myself, and I said yes. Every summer, I'd go to Red Auerbach basketball camp and the legendary coach would smile for pictures and supervise hordes of kids who thought that this was their tryout to become an NBA player themselves. Whenever I went to camp, I wore my Converse Weapons—the same ones that Bird endorsed. But for some reason, I still couldn't hit a jumper like #33.

I wore those shoes whenever Alan took me to the Garden as well, and when he came by one afternoon a little earlier than usual, he told me to get ready. He had a surprise for me. So, I grabbed a green shirt, laced up the Converse, and ran outside. *A surprise? What could it be?* No matter how many times I asked on the ride there, Alan wouldn't even give me a hint.

Outside, the Garden was quiet. It was still a ways away from tip-off, but Alan marched me to a special entrance. An employee stood outside, guarding the door.

"We're here for shootaround," Alan announced.

I had to fight myself to keep from letting out a squeal.

But that wasn't all. I assumed we'd sit in Alan's usual seats, watching the Celtics warm up from a safe remove. Then we were led through the underbelly of the Garden, through one nondescript room after another, until there it was: we stood ground level, right on the sideline, and I felt as if I floated when I took my first steps on the parquet court. Here came the Celtics. Parish ran by, even larger in real life than on my TV or up in Alan's seats. Ainge hit shot after shot. Auerbach watched it all, writing little notes to himself. Alan stood there next to me, beaming. I tried to take mental pictures so I could brag to all my friends at school. Then I realized that something wasn't right.

Larry Bird was nowhere to be seen.

"Is he hurt?" I asked Alan. I didn't need to specify who "he" was. Alan knew exactly who He was.

"I don't think so," he said, and then I started going through the scenarios in my head. *Oh no. There are only eight games left in the season. If we don't have Larry we might not get the top seed, and if we don't have home court advantage, then even Larry Bird might not be able to lead us to victory . . . unless—oh no. No no no. What if he missed the playoffs? Oh no no no no no no.*

Remember what I said about sports helping us imagine dream worlds? This was the flip side of that type of dreaming. In my head, this was doomsday.

Then the energy in the whole building shifted. In college, I had a professor who tried to explain Einstein's Theory of Relativity to our class of liberal arts majors (bear with me here). She asked us to imagine space-time as a trampoline. Something with a lot of mass, like a star, is the equivalent of a bowling ball, she said. Most everything else is like a marble. What happens when you put the bowling ball in the center of the trampoline? All the marbles fall toward the center. The bowling ball literally bends space-time.

The entire time she taught the lesson, I was thinking about that moment in the Garden. I was standing on the sideline when I noticed some movement in the tunnel. Turns out, Larry wasn't hurt. He was just late. And now, there he was: the most famous man in Boston, the one who everybody wanted to be, and he was jogging right toward me. Everything, at that moment, fell into him. Time bent. Larry was the bowling ball, and everyone else there—me, Alan, even Red Auerbach—was a marble. Maybe that's why they call players like him "stars."

Larry walked onto the court, looked around, and then we made eye contact from across the court. He pointed at me—right at me.

"Hey, kid," he said. "Want to rebound for me?"

I wish I could regale you with all the little details of this shooting session. I wish I remembered how many shots Larry Bird sank, whether I sent balls back to him via a bounce or chest pass. I wish I knew what he said to me when we were done.

To be honest, though, my memory is pretty sparse. I think I blacked out from excitement. In fact, I *know* I did.

What I do remember was the feeling. I couldn't have been more excited, shaking probably, but for some reason, the entire experience felt incredibly peaceful. It must have lasted thirty minutes, but it felt like hours. It was only me and Larry Bird, the only two people in the entire world, and it was my job to make sure he got his shots up. It was my job to get him warmed up so that the Celtics could win. Greatness was happening here, and it was my job to facilitate it. A lot of responsibility for a preteen, sure, but I was happy to do it.

When a god gives you a commandment, you listen.

It turns out Alan was not just a season ticket holder. His auto glass store was also a sponsor to the Celtics' philanthropic foundation, and as part of that, he'd bought this VIP access. If Larry was our god, Alan was the apostle who had made this miracle happen. I'd be forever grateful.

That night, as I was telling my parents about what happened, I remembered the stories my mom used to tell me and my sister as a kid, the legends she shared from India. I remembered Papa telling me of the moment he decided he wanted to go to medical school, reading that short story and feeling like the doctors could perform miracles just like those mythical heroes. Just for a moment, I started to get the first inklings that this whole love affair with the Celtics, and my experience with Larry Bird, might be about a lot more than just basketball.

////////

The best of those Celtics teams, though, came in 1986. Those guys had everything. Four Hall of Famers wore green and white that year: Larry Bird, Robert Parish, Kevin McHale, and Bill Walton. No team had their size—or their depth. They won 67 games, locking up the top seed in the Eastern Conference along the way. I told everyone I knew, "We're going to win another championship."

When the playoffs came, my Celtics obsession reached a fever pitch. Homework quickly became an afterthought on game days, and I would rush home to see if Papa came home from work with good news. "Gautama," he said one day, "Alan is taking you to the first two games of the playoffs."

My golden ticket entitled me to watch my Celtics play the Chicago Bulls, led by a second-year starter named Michael Jordan. But these weren't *those* Bulls teams yet. Coach Phil Jackson would not roam the sidelines for another year. Scottie Pippen still played for Central Arkansas. MJ broke his foot earlier in the year, and coming into the playoffs, he hadn't played a full game in months. Chicago finished the year 30-52—less than half the amount of wins as Boston.

This, I thought, *should be a walk in the park.*

The energy in the old Garden always registered as unique, whether preseason or rivalry games. But few things match the intensity of a playoff matchup. Even fewer can match the feeling that reverberates when something more than a championship is on the line—when everybody knows that their team has the chance to go into the record books. That's what I remember feeling entering the Garden before Game One. The 1986 Celtics could go down as one of the best NBA teams ever. I wouldn't just be watching a game. I would be witnessing greatness. And there were tens of thousands of others around me who knew the same thing.

Our home team did provide a glimpse of the squad that would go on to win a championship. But little did we know that the most incredible display we would witness would come from the second-year player from Chicago with the flashy sneakers.

Jordan seemed to hang in midair when he played. If Bird seemed to bend the laws of physics, Jordan simply defied them, floating above it all until he got his perfect shot. As an eleven-year-old, I couldn't understand how he was doing it. I asked Alan if he had ever seen anything like what MJ was doing. "No," he said. "But it's pretty amazing, isn't it?"

MJ scored 49 points in Game One, but the Celtics still won. I rambled the whole way home. Bird was amazing, and McHale was the best big man in the league. But Jordan? Nobody could do what he could do. I wondered how he could top himself in Game Two.

I didn't know this at the time—nobody would for several years—but on the off day between games, Celtics guard Danny Ainge golfed with Jordan. MJ was (and is) a serious golfer, and he knew that Ainge liked to swing a club, too. So, before the series started, the Bulls star called Ainge and asked for a recommendation: "Which eighteen should I play in Boston?" Ainge invited MJ to be his guest at Framingham Country Club. When the duo got there, everything that had been going right for Jordan on the basketball court started to go wrong on the links. He missed fairways, caught himself in sand traps, and couldn't sink a putt to save his life. Ainge clobbered him.

Michael Jordan, as you might have heard, does not like losing.

If he couldn't get even on the greens, he'd do it on the court. As the twosome left the course, Jordan paused for a moment. "Hey," he told Ainge. "Tell your boys that I've got something for them tomorrow."

Game Two was a Sunday, a school night, but my parents knew how much it meant to me. They told Alan to bring me home as soon as the game ended. But no matter how late I stayed up past my bedtime, it would have been impossible to get drowsy in the sold-out Garden that night. There was that growing anticipation, thousands of people gathered together to witness something special. I remember somebody in our section saying, "Jordan had his night last time. No way he can do it again."

I should pause here to remind you: this was before Michael Jordan was *Michael Jordan*. Please excuse our ignorance, the shear sacrilege of doubting the GOAT. We were about to learn our lesson.

The Bulls jumped out to a 4–2 lead after a few possessions, and didn't surrender their lead until late in the fourth quarter. As for MJ? Game One had been just a preview. He scored at will, dunking, slashing, shooting, and stealing the ball whenever he wanted. DJ, our best defensive player, tried to guard him. No luck. Bird didn't fare any better. Walton, who had six inches on MJ, even tried for a bit and grew so frustrated that he fouled out.

In all my years watching sports, I've still never witnessed a more dominant performance in person than the one Jordan notched that night. He could do anything. A strange thing happened, too. Everyone in the Garden that night wanted the Celtics to win, obviously, and as they chipped away at the Bulls' lead, our cheers grew louder and louder. But unlike those games against the 76ers, nobody actively rooted *against* MJ. Instead, this immense sense of respect filled the walls of that old stadium. We were there to see greatness, and even if it didn't arrive in the form we expected, it was hard to feel anything but awe for what we were witnessing.

Late in the fourth quarter, Bird hit a three as the shot clock expired, and the Celtics had their first lead of the game. I high-fived everyone sitting near me, but at no point did I feel confident. I knew, as everyone else did, that we couldn't guard Jordan. It was going to come down to whoever had the ball last.

That's exactly what happened. And luckily for us, after two overtimes, the Celtics put up the last shot and won. Jordan, when it was all said and done, scored 63 points—an NBA playoff record.

Alan and I didn't talk much about the Celtics on the way home. How could we discuss anything other than MJ? "I swear," I told him, "it seemed like he flew."

"I know," Alan said.

The next morning as I woke up extra groggy for a Monday of school, my mom handed me the *Boston Globe* sports section alongside my breakfast. There was a huge photo of MJ, and in the story recapping the game, I read a quote from Larry Bird. "That wasn't a basketball player," he said. "It's just God disguised as Michael Jordan."

I suppose I knew that Jordan wasn't an actual deity. But reading that statement—from Larry Bird, of all people—rang as true to me as anything I have ever heard in any class, book, movie, temple . . . anywhere. That *wasn't* a basketball player I watched the night before. He had an aura to him, a sense of invincibility and inevitability. Everyone in Boston Garden knew it. Everybody felt it. Everybody watched with reverence, and if you polled Celtics fans as they left their seats that night, I guarantee you a vast majority would say that they would believe it if you told them that Jordan could walk on water. After all, we had just watched him sail through the air.

These were heady thoughts for an eleven-year-old, especially one from as spiritual a home as mine. But all of the stories and lessons my parents told me felt far away and inaccessible. I assumed that they were true, but who knew? They were just stories. Here was something I saw with my own eyes. Michael Jordan *was* a god—at least in some way.

It would take me a while to work out all the details. But I converted right there at the breakfast table. I believed in the Religion of Sports.

LARRY BIRD

NICKNAME: Larry Legend

DESCRIPTION: Forward, Boston Celtics, 1979–92, Basketball Hall of Fame

FACT #1:

One of the greatest pure shooters in basketball history, Bird led some of the NBA's most dominant teams, winning championships in 1981, 1984, and 1986.

FACT #2:

Hails from French Lick, Indiana, population: 2,059. Bird was such a popular player that attendance to his final high school game was nearly double the amount of people who lived in French Lick, with about four thousand fans cheering for him.

WISDOM:

"A winner is someone who recognizes his God-given talents, works his tail off to develop them into skills, and uses these skills to accomplish his goals." *—Larry Bird*

CHAPTER 2

True Believers

Many believe in the Religion of Sports. Some people live it.

I've used sports as a metaphor for so much of my life. When you think about [the Patriots' motto] "do your job," how does that apply to life? Well, be the best husband you can be. Be the best father you can be. Be the best at your job you can be.

—Tom Brady

ONE DAY, IN the seventh century of what today makes up Saudi Arabia, a group of men approached the prophet Muhammad. The Prophet had grown to be one of the most powerful figures in the region, spreading the word of Allah while leading an accomplished army. The Islamic Empire slowly expanded, and soon pagan idols fell throughout Arabia. As the religion spread, more and more people started to convert. Some of them believed in the Quran. But others just viewed their professed belief as a token they could exchange for protection and wealth. Maybe they, too, could share in the spoils of war.

The men who approached Muhammad on that day fell into this latter group. They claimed to follow the religion, but they didn't practice its beliefs in earnest. They only went through the motions of praying and visiting mosques, but as soon as those obligations were completed, they would live their lives according to a different type of moral code. Now, meeting the Prophet, one of the members of the group spoke. "We," he declared, "are Muslims."

Muhammad must have sensed their apathy. They were Muslims in name only. "You have not yet believed," he told the group, "for faith has not yet entered your hearts."

The difference between someone who claims to follow the Quran and someone who is a true believer is an important distinction in Islamic faith. There's a special term for people who are true believers: *mu'min*. *Muslim* is the generic term for anyone who practices the faith. But a *mu'min* is someone who believes in the faith, deep in their heart. He or she is a person who has no more doubts about their belief in Allah. Muslims read the Quran. *Mu'min* live it.

That dichotomy, between Muslim and *mu'min*, reminds me of some of the best athletes in the world. The difference between the greatest players and the ones who are just good enough has been scrutinized for decades. There are theories and philosophies, books and dissertations written on the topic. One thread inarguably connects all these high performers, though, and that is that the greatest competitors are true believers in the Religion of Sports.

Here's an example: What is the thing that separated Wayne Gretzky from every other hockey-obsessed kid in Canada? We can start to find the beginnings of an answer by considering how he approached the game. When Gretzky was young, he sat in front of the TV watching NHL games every night during the season. In his lap, he held a blank piece of paper and a pencil, and as he watched, he'd slowly start to trace. He wouldn't look down at the doodle, just at the screen, but as the puck moved, his pencil followed. He was charting the path of the puck, for the entire game, with constant loops and zigzags. At the end of each period, Gretzky would look down and study the movement.

He knew that the darkest portions of the page revealed the places that the puck spent the most time.

It's no surprise, then, that when Gretzky grew up, he had the ability, as many commentators have noted, to somehow skate to the exact same place where the puck was heading. Of course Gretzky could sense where the puck would be, because Gretzky wasn't just a believer in hockey—he was a *mu'min*. He didn't just play hockey. He lived hockey. Wayne Gretzky was a true believer.

Work ethic, talent, and luck are all just the by-products of an unwavering belief in one's sport. It's what happens when somebody has dedicated themselves fully to their sport. It's what happens when practice and study becomes a way of life. And it's exactly the path that Tom Brady followed to become the greatest quarterback of all time.

/////////

Tom and I were introduced by a mutual friend sometime around the 2013 off-season. After the season, Tom spent his time in Los Angeles, not far from where I live, and one day, he invited me to visit at his home. As a Patriots fan, it was like being summoned to the Vatican to meet the pope; for years, watching Tom Brady made me believe in miracles. He brought me and my friends closer through every Super Bowl run. Now I was getting to be face-to-face with him? I spent the whole drive there trying to figure out which Super Bowl I was going to ask him about first.

But when I arrived, Tom was the one asking me questions. How did I get to be a Patriots fan? Did I think the Seattle Seahawks could repeat as Super Bowl champs? And finally, what projects was I working on? At the time, I was pitching a series based around this idea of a "Religion of Sports," and so I explained the concept to him and showed him a visual reel we'd created that captured the feeling we were trying to evoke. In further explaining the concept to him, I used his own life as an example: the Patriots' Way was his dogma, Sunday games his church, the Super Bowl his annual pilgrimage.

Tom nodded. He told me of growing up in San Francisco watching Joe Montana and how he really did worship #16. Tom was in the crowd at Candlestick Park the day of "The Catch," all of five years old, and he recalled how the 49ers victory felt like a miracle to him back then—and now. "I was in awe," he told me. Tom said that still, whenever he sees Joe Montana, he gets flustered. Being on the same stage as his childhood hero never gets old—even for the greatest of all time.

We chatted some more, and then I said goodbye. The next day, Tom texted me. "I was thinking some more about your Religion of Sports idea," he said. He told me he loved the video I'd shown him. "I get it. We should talk about it some more." And so, we did. Eventually, along with NFL Hall of Famer Michael Strahan, Tom came aboard the Religion of Sports television series as a producer. In truth, I'm pretty sure he didn't know what that role meant, but he wanted to help and trusted me and Michael. We sold the show and started to make it a few months later under the auspices of one simple idea: "why sports matter." As time passed, I reminded myself that the "Religion of Sports" was more than a television series. It was a big idea, and I wanted to do more with it. So, I convinced Tom and Michael to become business partners; we were going to produce even more content that explained "why sports matter." To that end, I pitched Tom for years on letting me make a documentary following his process and preparation, but he always politely told me, "Not yet."

Then, one day, he was ready.

Tom texted me one March afternoon in 2017, just a few weeks after he won Super Bowl LI against the Atlanta Falcons. His fortieth birthday was fast approaching that summer, and an idea had been forming in Tom's head. He was going to undergo his normal off-season training program, and then see if by his fortieth birthday (in

August), he could beat his 40-yard dash time from the NFL Combine (5.28 seconds when he was twenty-two years old). "Let's film the training," he wrote. "It'll be fun."

Even though I wasn't entirely sure what we were going to be shooting or where it would end up, who was I to say no? I got on a plane and flew to Boston, where Tom was based. And just like that—about three weeks after completing the largest comeback in Super Bowl history—Tom was in his backyard in Brookline, Massachusetts, throwing around a football in full pads, using resistance bands for strength training, and running wind sprints while his dog chased him back and forth on the lawn.

Over the next few weeks and months, we kept on it, filming a variety of different workouts and training sessions. One time, when we were back in Los Angeles, Tom summoned me to a local middle school where he was working on his mechanics with former Major League Baseball pitcher and quarterback guru Tom House, throwing to Patriots receiver Julian Edelman. "Elite quarterbacks don't come in to get five percent better," House explained to me. "They come in to get one percent better."

But Tom wasn't even trying to get *one* percent better. It was more minuscule than that. Tom obsessed over the tiniest of improvements, attempting to "rewire his nervous system," in his words. House described the adjustments as splitting "frog hairs."

Brady wore sweats, a red TB12 shirt, with his pads on top and nothing over them. He wore his silver helmet, too, although all its stickers had been removed. The mechanics of throwing with pads and a helmet are just slightly different—and so Tom sought out the most authentic experience possible.

Tom noticed that before he released the ball, he was overextending his left arm by just a few inches. That overextension caused his head to come slightly off level, and his accuracy to suffer just a hair. But again, being even a hair off isn't good enough. For Tom Brady, the point of practice is to split those frog hairs. Did I mention this was a guy who at that point had been in the league for eighteen years, gone to seven Super Bowls, and was already widely considered the GOAT?

"I'm sorry I'm getting on you," House said at one point after critiquing Tom's release.

"No," Tom said. "It's what I need."

Tom is uncompromising in these workouts, and after each throw that looked perfect to someone like me, he'd kick himself and ask to do one more. For Tom, football is craft, and the way he attacks that craft is something that I've found myself using as inspiration in my own life. Tom famously told reporters that he wanted to play until he's forty-five. He told me the same thing (and also occasionally mentioned the big 5-0). He retired and then unretired just a few months later, and now he's retired again. That desire to keep playing doesn't come from a place of hubris. It comes because he felt like at the end of his career, he was as close as he'd ever been—maybe even as close as anyone has ever been—to finally mastering his craft.

This footage (and much more) ended up coming together in a documentary series Tom and my creative team at Religion of Sports produced for Facebook called *Tom vs. Time*—although it shifted from documenting a race between forty-year-old Tom and his twenty-two-year-old ghost into a more introspective look at what it means to dedicate your life to a single pursuit.

"What motivates me is I coulda, shoulda done better," Tom said in 2017. "I think, *Man, I'm in my eighteenth year. I've had all this knowledge, played so many games, so many experiences, so many teammates, so many game plans.* Like, I should be—could be—perfect."

Perfection is what Brady is chasing. When he loses, he's tormented. "I think about losing for days," he says. "For days. You wake up thinking, *Oh my God.* It's too painful. And it could feel like just a game, but because it's more than just a game to me, it feels like I'm losing in pursuit of what my life is. Thank God there's another game the next Sunday."

If he wins, there's always something more to be achieved. "The joys of winning are great," Tom explained. "But I sweep those under the rug so fast that there's only a brief moment in time where I enjoy the experience of winning, and then I'm thinking about the next game. And I'm thinking about winning again."

He didn't always think that way. Tom told me a story once about Super Bowl XLVI, when he lost to the New York Giants in 2012. It came in the midst of a rare, challenging period by the Patriots' standard. They had lost to the Giants in the Super Bowl just a few years earlier, which spoiled an undefeated season, and this loss marked seven years that the Patriots had gone without winning a championship. Not a lot of time for most franchises. But for the Brady-era Pats? It felt about as long as the Red Sox's eighty-six-year drought.

Tom couldn't sleep the night after the game, tossing and turning in his Indianapolis hotel room, rewinding through certain plays in his head. What could he have done better? What went wrong? Was he ever going to win again?

And then, Tom realized something. It became a sort of mantra that he's carried with him ever since: "What happened yesterday is the past. The only thing you take from it is the lessons, the things you learned from it."

That comment seemed to unlock something in Tom, and keeping things in perspective became a priority. You might even say that he learned to value losses as much as wins. After a slow start to the 2018 season, he once told me, "If you're 2-2 and haven't learned anything, then frankly, you have a problem." Maybe it's no surprise that a few months after the 2-2 start, Brady led his Patriots to appear in another Super Bowl.

If it sounds like Tom Brady is never satisfied, that's not exactly a fair representation. He loves the pursuit. The losses are torturous and the victories are fleeting, but the process gives him purpose. "His ability to play with joy on any given Sunday is really based on his ability to find joy in the week leading up to the game," his longtime trainer, Alex Guerrero, once told me. Chasing that perfect throw is a 24/7 operation for him. He finds comfort in that quest.

When Tom goes on vacation—to Costa Rica, Montana, China—he brings an oversize duffel bag filled with cones and bands and balls, and without fail, he wakes up at the crack of dawn and finds a field to practice. Football, for Tom, is so much more than just a sport. "I am trying to find that deeper purpose," he said. "To live it through sports in a very authentic way makes so much sense to me. Having these dreams and goals and aspirations, and waking up and putting in the work and miracles happening and all this magic that sports create. I'm in the middle of it. I get to live that through sports." Later in that conversation, Tom told me that football "allows me to be who I am in a very authentic way that is hard for me to be when I walk off the field."

And so Tom finds fields like that one in LA (or on a beach in Doha or a park in Tokyo or a gym in Beijing—all places I've traveled with him over the years) and tries to make the tiniest of improvements. He threw for hours that day in LA, and it was easy to imagine him as a JV backup doing the same thing with his high school friends, each throw building to the next until that day, when he's surrounded by a camera crew and is right on the edge of perfection. "I'm starting to feel it," Tom said, and he'd ask Edelman to run another route. Just a little off. "Run it again, Jules." Shoulders a touch out of position. Again. A go route this time. Brady fired his hips, and the whole mechanism uncoiled in perfect harmony. As Tom finished his follow-through, he paused for a moment to remember the feeling, so that he could replicate it on the field next year.

"That's perfect," he said. Then he asked Jules to run it one more time.

//////////

In their approaches to their respective sports, there are few athletes more different than Tom Brady and Ultimate Fighting Championship (UFC) superstar Conor McGregor. McGregor is loud and vicious and visceral. But in a way you might not expect, Conor and Tom remind me of each other. No matter what's happened in their careers, they've continued to dedicate their lives to their sports.

During the middle of the pandemic, in late 2020, Conor invited me and my team to follow him in the run-up to his second fight against Dustin Poirier at UFC 257. I've always been fascinated by the UFC, and tracking its rise over the last three decades has felt like watching the birth of a new sect of the Religion of Sports, led by its larger-than-life leader, Dana White. Of course, UFC is a show and a spectacle, but it's also primordial. "Nothing quite feels like losing a fight, because it's primal," said Colin Byrne, Conor's trainer. "If the referee wasn't there, you would've died. And you know it."

You can have the best trainers in the world along with the greatest training equipment, but at the penultimate moment, your team hands you off to walk into that cage, and they close the door behind you. From then on, it's just you and whatever physical, mental, emotional, or even spiritual grit you have to face the person in front of you. There's no way to scheme your way out of it. It's raw and violent, sure, but there's also something deeply honest about the sport. Only one fighter walks out.

With his antics, ego, and violence, Conor McGregor is the easiest guy in the world to root against (and he's certainly a huge challenge to deal with). As Dana White described him to me, he's just an absolute "savage." But Conor also might be the person I most admire spiritually. Like Tom, Conor's sport is not just *a* way of life to him. It's *the only* way of life.

McGregor's fight against Poirier would be held in Dubai, at so-called Fight Island. Much of his training, though, occurred in Lagos, Portugal, at a no-frills mixed martial arts (MMA) gym near the coast. These workouts were nothing like Tom's New Age pliability routines and more like how Muhammad Ali, one of sports' and Islam's great believers, utilized his famous Deer Lake, Pennsylvania, training camp. Ali nicknamed that place "Fighter's Heaven."

Lagos, like Deer Lake, stripped training down to its essentials—no fancy decorations or unnecessary accessories. Conor sparred in an octagon, squatted barbells, and ran up and down beachside hills over and over again. One day, when Conor was practicing kicks, he hit the pad so hard that it started to tear. If anyone ever asked—and even when they didn't—Conor would say he was feeling like he was in the best shape of his life. His team echoed the sentiment. When he boarded his private plane to Dubai, Conor said he knew he was going to win.

Conor is many things—a showman, a villain, and probably crazy. He literally fought his way from nothing to become iconic, or, in his words, "notorious." Before making it to UFC, Conor got by on a €188-per-week welfare check. He found work in a factory and later drove a cab. In 2015, after signing a record-breaking contract, he bought his entire family matching BMWs and paid off their mortgage. Conor wears tailored custom suits in loud patterns and when not in formal wear, he typically doesn't wear a shirt at all. He drives Rolls-Royces. In Dubai, he chartered a massive yacht that included not one, not two, but three slushy machines, because his children love them.

The point is, Conor has all the fame and fortune someone could ever imagine, and his legacy as one of MMA's fiercest fighters ever remains intact. There's no need to keep getting into the octagon, no need to risk injury or defeat. No need to keep running up a hill on a

Portuguese beach over and over again. But Conor can't—or won't—give up. He's a gladiator, a scrapper. Most of all, he's a fighter.

In the moments before the Poirier fight in Dubai, Conor sat with his eyes closed, meditating in a crisp dress shirt while getting his wrists taped. That feels like as good an image as any to describe the man: showing the world who he is with the fancy shirt, being locked in mentally, and all the while preparing to fight. "My weight's good," he told me before the fight. "My body's good. But my mind is best. Bulletproof."

Yet when Conor got in the octagon and that door closed behind him, Poirier's vicious kick kept striking Conor on the shin. Before long, Conor could barely feel his own leg. He limped around, still dodging and throwing punches, but being one-legged isn't a recipe to win a bout in UFC. Poirier knocked Conor straight in the face, causing him to stumble backward, and then Poirier pounced on top of him. Before he could do any more damage to Conor, the referee waved him away. Conor had been defeated by technical knockout.

He limped back to the locker room where our camera and crew waited. Conor slunk into a chair against a wall, saying nothing except to wince with pain and grab at his leg. His whole team sat around the room, but nobody spoke. All eyes were on Conor. Someone handed him water; someone else grabbed an ice pack. Finally, Conor put his head in his hands, then pulled himself back, sitting straight up to take a deep breath. He cleared his throat. "That's that," he said. "That's it. That's it."

With that, Conor had processed the loss. He'd moved forward. Pulled himself off the mat, ready to fight again. That was all there was to it. As soon as he got healthy, Conor found his way back to a gym. Not because he needed the money. Not because his ego demanded that he avenge himself. Not because it would be easy.

Because he's a fighter.

And because, like Tom Brady, he's a believer.

It's his way of life.

//////////

Ask anybody who knows Tom Brady, and they'll tell you that he only has two priorities in his life: family and football. Everything is about family and football. When he's not playing football, he's with his family. When he's not with his family, he's studying or preparing for football. He's not the type of guy to go and grab a beer with some buddies, nor is he going to be caught by the paparazzi at the hippest restaurant. It's football and family—that's it.

Tom talks often about "inputs and outputs." He believes that in order to maximize what you get out of something, you have to control what you put into it. "If you want to perform at the highest level, then you have to prepare at the highest level," he says. That partially explains his famously strict diet. Tom views it as his job to protect his body, and so if he's going to get the most out of it, then shouldn't he only put the best into it as well? He's relentless about preparation. Even at the end of his career, when you might expect the oldest player in the league to struggle to make it through the grind of a seventeen-game season, the physical aspects of the game come easy. "It's effortless," Tom says, "because it's just so synonymous with my being."

Tom can read any defense and make any throw, but the challenge to him remains the mental game. Tom believes that football is nothing but a game built to measure someone's mental toughness. As a result, he's maniacal about crafting the best possible inputs to get the best possible outputs out of the mental part of his life. No distractions, then. It's just family and football, family and football, family and football.

He only has so much time to be with his kids. He only had so much time to play in the NFL. Why waste any of it?

//////

While he played, this was Tom Brady's schedule from August to February for twenty-two years:

MONDAY: Watch film
TUESDAY: Watch film
WEDNESDAY: Practice
THURSDAY: Practice
FRIDAY: Practice, watch film after practice late into the
    evening
SATURDAY: Film
SUNDAY: Film in the morning before the game. Then,
    bodywork after the game.

Oh, and his day started each morning at 5 a.m. with a specially made smoothie. Tom followed this schedule like clockwork, and so when he was in the middle of a season, the days all blended into each other. This was hard work—and a lot of work—but Tom loved it.

And I mean *loved* it. He lived it. More so than any athlete I've spent time with, Tom Brady found joy in even the most tedious aspects of his craft. Kobe Bryant loved basketball because it was a space where he could channel his emotions and vent. Simone Biles finds purpose in pushing herself to be the best athlete possible. Steph Curry uses three-pointers to calm his mind. But Tom just found every part of football to be thoroughly fun.

"In the end," Tom explained to me during the 2017 season, "my life revolves around football. It always has been, it always will be, as long as I'm playing. I've given my body, my everything—every bit of energy for eighteen years to it. So, if you're going to compete against me, you'd better be willing to give up your life. Because I've given up mine."

If you want to appreciate the way that Tom views football, you have to observe the way he dissects game film. His home office is his laboratory. The desk looks fairly typical, decorated with knickknacks; this isn't a man cave–type shrine to all things NFL. There's a globe, a small figurine of frogs jumping, and framed photos of his family. On the table behind the desk is a Lombardi Trophy.

There's a cabinet right under his printer, and on a shelf inside it sits a row of thick navy blue binders. On the spine of each one is a date: 2000 . . . 2001 . . . 2002 . . . They're stuffed with photos, clippings, and other memorabilia from every season he's played in the NFL. One day while I visited, Tom pulled out his tome on the 2016 season. He wanted to show me something, and he held up a piece of paper: the letter from NFL commissioner Roger Goodell detailing his suspension due to Deflategate. "Just a nice way to remember," Tom said. The binders are also full of play charts and scouting reports from every game he's ever played. Sitting in his office that day, Tom was a few months removed from leading the Patriots to a stunning comeback to beat the Atlanta Falcons and win another Super Bowl, and he pulled out that season's binder, too. He showed me the agenda for the Patriots' pregame team meeting before the Super Bowl. Tom had covered every inch of the paper in notes, sometimes transcribing nuggets of wisdom from Coach Bill Belichick: "Prepare and play well." . . . "Super Bowl environment is all about hype and ridiculous bullshit that will go on." . . . "It's a great week. It's about competition. The team that wins is the one that works the hardest."

Why does he keep all of this? Is he just nostalgic? Is he a hoarder? Tom views these mementos as a unique competitive advantage. Before he dives in to a new game, or a new season's challenge, he'll recall a time in his career when he confronted a similar situation—maybe a game against a particular defensive coordinator or a time when the outside noise felt deafening—and consult his archive. Tom will read these notes then to see how it compares with the way he remembers things and if there are any useful lessons to be learned. Then he'll watch the film. It always comes back to film.

"I could literally, like, watch film all day," he told me once. "It's almost like soothing . . . I can just go four, five hours without getting up from this chair."

He really will. Tom sank into his tan leather office chair, queued up game film on his laptop (a team-issued Dell), and connected a small remote, which he held in his right hand, thumb hovering over the rewind button. If he's not clicking anything, the footage will play. Tom would let the action move forward no more than a second before going back to the beginning. He'd spend five minutes watching the very first second of each play after the ball is snapped. He looked at the defenders (What were they doing before the play?), he studied his footwork, he identified potential holes in the defense's coverage. "When you've been doing something for so long, you don't just see tape," he said. "You see the depth of everything." As Tom watched, he made notes, and every now and then, he'd call his coaches to share his thoughts. I was with him once in New England before he played the Cincinnati Bengals in 2017. He was scrutinizing one of Cincinnati's preseason games when he noticed a gap in their coverage. He called Josh McDaniels, the Patriots' offensive coordinator, and left a message. "Hey, on the 96 Cavalier Roy, I was watching Cincy preseason. After Gronk hits the end, he can go vertical. . . . Cincy preseason play 30. Check out that play. Thanks." McDaniels called him back a little while later. He had updated the playbook.

Over time, Tom has come to master the game better than anybody. When he gets on the field, as a result of all his experience and all the time he spent studying his opponent, he often knows exactly what every defender is going to do. The first thing he reads is the defense's formation. Basic stuff: *Who's standing where?* From there, he goes deeper. He studies each player lining up across from him, looks at their eyes and how their weight is distributed. He's looking for tells, like a poker player trying to call a bluff. "The game for me is very calculated," he said. "It's very strategic. It's much more like chess than checkers."

When you start to push deeper with Tom, when you ask about his philosophy, or when you watch him work, you start to see that the way he characterizes his obsession—that he has "given up" his life for it—is not totally fair. Football hasn't forced Tom to sacrifice his life. On the contrary: football is the thing that has *given* Tom the life he's led since he was a child. He believes in it on a level that can only be described as spiritual. And he's learned from it in a way that has echoes in religion as well. He approaches each game, each season, as a lesson. Then Tom uses what he learned off the field to be a better person off of it.

Here's one example that Tom shared with me while we were working on our docuseries *Man in the Arena*, which I often refer to as *The Tao of Tom*. Before the 2004 season, Tom didn't have the same balance in his life that he does today. He was young, rich, and famous, and so he enjoyed being a celebrity, making appearances at the hottest parties and in the tabloids. He even put on a little weight. One day, linebacker Willie McGinest pulled the quarterback aside. Willie said that if Tom wanted the rest of the team to hold themselves to a higher standard, then Tom had to hold himself to one, too.

And thus "The Edgers" were born.

That's what Tom, Willie, and several other Patriots leaders, including Tedy Bruschi, called themselves as they pushed each other to work harder than ever before—as they always hunted for a hidden edge. "You weren't held accountable by the coach," said Brady. "You weren't held accountable by the fans. You were held accountable by the guy that was sitting next to you every day." If you showed up to lift weights at six thirty in the morning, somebody else would be there at six. If you studied film for one hour after practice, somebody else would study for three. "We outworked you," said Brady. "We outcompeted you. And then, when the chance came, we outwilled you."

The rest of the Edgers have retired now, but Brady still brought that philosophy to the training facility every day in Tampa for the Buccaneers. He also brings it to every aspect of his life, especially his role as a father. Being around Tom has made me notice the ways that I can find an edge in my own life. He's shown me that leadership isn't just about giving motivational speeches. It's about finding edges, about always remaining aware of what lesson you can take from each opportunity.

I love the Edgers' example, because it's so classic Tom: taking something he learned on the football field and blowing it out into a full-blown life's philosophy.

//////

The Friday night before Super Bowl LIII against the Los Angeles Rams (the Patriots won), Tom invited me to his hotel room—no cameras, just an opportunity to catch up. He was stretching, getting a massage, and preparing for the game while we had a conversation about the match-up, about his mindset, about what another championship would mean.

Then Tom started telling a story.

He said that there was once a farmer in China whose horse ran away. Neighbors came from all over to share their sympathies with the farmer; it was his only horse, after all. "That's horrible that your horse ran away," the neighbors said. The farmer responded with a single word: "Maybe."

The next day, the horse came back—and it brought ten other wild horses with it. The neighbors followed close behind. "How lucky that you have ten new horses!" they said, and once again, the farmer said only, "Maybe."

Tom kept going. The farmer's son woke up early the next morning to break in the new horses. He made steady progress, and then one of the mustangs bucked him. When he fell, he broke his leg, and the neighbors rushed back to the farm. "What a terrible thing to happen to your boy," they said. The farmer replied, "Maybe."

I looked around the room to see if anybody else was hearing what I was. *Is the greatest quarterback of all time seriously reciting a fable to psych himself up before the Super Bowl?*

A day later, a war broke out. Officers marched from door to door, gathering every able-bodied young man to serve in the army. When they reached the farmer's house, they asked about his son—only to notice that with his broken leg, he couldn't serve. When the army marched off to battle, the remaining neighbors approached the farmer. "What a blessing," they said. *Maybe.*

The story floored me then, and it still floors me now. So much so, in fact, that I brought it up with Tom again a few years later. This time I asked some follow-up questions. What did that story mean to him? Why was it so memorable?

"We don't have the perspective of what's going to happen in the future," he explained. "And when we don't have the perspective, we don't understand whether what happened was good or bad. We just have to understand that there's a lot of things at work, and what we may think is good may not be good, and what we think may be bad may not be bad—because the future will tell."

You can encode so much of the Patriots'—and Brady's—story onto the fable of the Chinese farmer, but perhaps most of all, it reminded me of the loss to the Giants in Super Bowl XLII, when the Patriots suffered their only defeat of the season. After the game, as Tom walked to the team bus, rain poured down. Phoenix averages nine inches of rain a year, but it poured that night. "We're sitting there on the bus in the pitch black," remembered Tom. "Everyone is just sitting there dead quiet. Our hearts were broken."

Tom told me about the way that he felt in the weeks that followed, how he couldn't sleep, how he took responsibility for the loss. Then he started to imagine a different outcome.

"Had we won that game, I don't know," Tom said, pausing. "I'm not a big hypothetical guy, but maybe the desire is a little bit different. If you're looking at a silver lining, maybe the desire to reach that point, maybe I would've been fulfilled not to stop playing at that time, but I don't know. Maybe I'd play another seven or eight years and I'm fulfilled."

Maybe.

//////

As we neared the end of the season we spent together filming *Tom vs Time*, the Patriots found themselves deep in the playoffs once again. I met Tom in Boston for the AFC Championship game against the Jacksonville Jaguars. All week long, Tom's hand was the talk of the sports world. He'd sliced it open in a freak practice injury while handing the ball to a running back, and now nobody was sure whether or not he would be able to perform against the Jaguars' defense, which was ranked first in the league that season.

We spoke on the phone the evening after the injury. For the first time all year, he seemed a little unnerved. "This season can't end on a handoff," he said. "I didn't come this far for it to all end on a *handoff*." He had twelve stitches on his throwing hand. Tom said that he wasn't sure if he was going to be able to play well. But he'd definitely be able to play. Tom Brady wasn't going to miss a playoff game on account of a busted thumb.

I met Tom at his house early in the morning, where he was cooking pancakes with his kids. The hand was bandaged up, and on the way to the stadium, stopped at a red light, he asked me, "What do you think?" and held up his hand. Even Tom Brady had some doubts that morning.

Tom was slow in the first half. The whole team was, down 14–3. Worst of all, star tight end Rob Gronkowski took a gnarly hit and left the game with a concussion. Heading into the fourth quarter, the Pats trailed 20–10, and even from my view up in the Brady family box, it felt all but inevitable that the season would end there. The Patriots' Super Bowl dreams crushed by . . . the Jacksonville Jaguars. As a fan, it felt humiliating. And as a storyteller creating a documentary series about this season, it didn't feel like a particularly satisfying ending.

Then, a miracle. Brady drove the Patriots down the field—85 yards in eight plays. On the Jags' next possession, the Patriots defense came up with a huge stop. And finally came another methodical, classic Tom Brady drive that ended with Tom finding wide receiver Danny Amendola in the back of the end zone for a circus catch. 24–20. Patriots win. They were going to another Super Bowl, Tom's eighth.

On field, the Patriots celebrated the AFC Championship, lifting one more trophy. After leaving the field and getting dressed, Brady met his family and friends down in the bowels of the stadium in a private reception room. Everyone cheered. He smiled. "We had 'em all the way," he laughed.

Characters lingered for about an hour until it was just me and Tom and his family. It was time for me to go, so I said good night. I told them, "Don't worry about me. I'll get a ride."

But the Bradys wouldn't have it. "What do you mean?" they said. "Of course you're coming with us." So I piled into the backseat of Tom's truck and let the camera roll as he drove back to his home in nearby Brookline. Occasionally he'd shake his head in disbelief and share an anecdote from the huddle with his family.

At one point, the Bradys dialed Gronk to see how he was feeling. He sounded a little groggy, his words slurring a little bit. "Playoff Danny," he exclaimed excitedly, remarking on Danny Amendola's now-heralded ability to show up big in the biggest games. Tom gushed. Gronk's energy was rising. He laughed: "We're goin' to the Super Bowl!"

We got to Tom's home super late—past midnight, and soon, Tom and I found ourselves alone in the kitchen.

Over time, especially that year filming our series, Tom had become a friend, but that doesn't mean that he ever stopped surprising me. I still was a die-hard Patriots fan, and I'd just witnessed firsthand one of the greatest performances in Patriots and Tom Brady lore—like Michael Jordan 63 points in double overtime. It hit me all at once that this wasn't a normal place to find yourself, sitting across from the greatest quarterback in football history after yet another memorable game. It felt like the end of an incredible journey that you only ever read about it in, well . . . a book like this. I'd gone from being a fan from afar to being as close as you could possibly get. The camera was off. We were just two believers in the Religion of Sports: one of us living it as a fan, the other as an athlete.

I asked Tom, "How did you do it?"

He shook his head. "I don't know."

I waited a beat. Then I asked again, "When did you think you were going to win?"

"Never," he said, and a smile crept across his face. "I mean, Gronk's out of the game. I thought we were screwed. But you have to think one possession, one play at a time. Can we get a first down? Late in the game, I'm just thinking that same way, every play, till it's over and you stop and think—" Tom's eyes were wide now. He waved his hands. "What just happened? That was a miracle."

He let it sink in for a minute. I didn't say anything, thinking about that, too. "Now that it's over, I'm thinking about it just like you are. I'm in disbelief." He sat in that surprise for a little longer before continuing: "Aaron Rodgers, he's a great quarterback, a *great* one, and he's been to one Super Bowl. Ben Roethlisberger, that guy's built like a tree. He's got a big arm. You can't tackle him—and he's been to two of them. Now here I am going to my eighth one. I don't say that to be conceited. It's just ... true. It's hard to believe. It's really hard to believe that it's true, you know?"

We talked a little while longer in the kitchen. Time passed, Gisele called from upstairs. "Tom, it's late. The kids are going to be up early."

I said good night and headed for the door. Then, Tom remembered I didn't have a ride. He told me that he didn't want me waiting outside so late for an Uber. I told him it really wasn't a big deal, I'd figure it out. But Tom wouldn't have it.

He tossed me the keys to his car. "Take the Raptor," he said, and he walked upstairs to go to bed. He'd be awake a few hours later, coming back down those stairs and into his office. He needed to get up *before* the kids did to start watching film and prepare for another Super Bowl.

Run it again.

TOM BRADY

NICKNAME: Tom Terrific

DESCRIPTION: QB, New England Patriots, 2000–19; Tampa Bay Buccaneers, 2020–22

FACT #1:

The undisputed greatest quarterback of all time, Brady has seven Super Bowl victories—more than any *franchise* in NFL history.

FACT #2:

In his nearly three-decade NFL career, he never had a single losing season as a starter.

WISDOM:

"There are a lot of guys who are all talk. They say they want to work harder and be the best, but they never pay the price. I love paying the price." *—Tom Brady*

CHAPTER 3

Myths

Myths provide the foundation for which a group can start
to believe in something larger than themselves.

It's entertainment plus fate, and if it exists for any reason at all,
it exists as fate's laboratory, a place where under controlled
circumstances we receive impossible data about the relationship between the accidental and the inevitable.

—Tom Junod, on sports

BEFORE IT BECAME synonymous with 26.2 miles, the word *mara-thon* was best known for its military significance. Along with the Athenians' defeat at Syracuse, the Battle of Marathon in 490 BC was one of the most consequential campaigns in the history of the ancient world.

Athens, at the time, was a relatively small city-state just two decades into a great experiment with a new form of government they called democracy. Persia was a mighty empire with a huge army, and its king planned to invade Athens, take over Greece, and expand his empire. It sounds like a pretty typical, long-ago war that you learned and then forgot in history class. But the Battle of Marathon had about as big stakes as a battle could have. If the Greeks won, democracy could continue to develop. If the Persians won, many of the Greeks' great contributions—to arts, science, philosophy, literature—would be crushed by totalitarian rule. Wrote the historian Sir Edward Shepherd Creasy in his 1851 book titled, humbly, *The Fifteen Decisive Battles of the World*, "On the result of their deliberations depended, not merely the fate of two armies, but the whole future progress of civilisation."

The only problem for the Athenians? The Persians outnumbered their forces three men to one. The Athenians needed backup, and so they turned to a special type of soldier called a *hemerodromos*, a day runner who specialized in carrying messages across long distances— quick and by foot. The messenger's name was Pheidippides, and the Athenians sent him to Sparta to bring back an army. Sparta, however, is 150 miles from Athens. And remember, time was of the essence.

Pheidippides ran. It was what he was trained to do. He jogged either barefoot or in sandals, with light armor. He ate figs, olives, and dried meats. He barely slept. Just two days after leaving Athens, Phe- idippides arrived in Sparta. He told of the coming attack, how Persia wanted to take over all of Greece, and the Spartans agreed to join the fight. There was one problem, though: due to the Spartans' beliefs, they could only join a battle when the moon was full—about a week away. Pheidippides needed to tell his army that they would have to wait for reinforcements.

So what did he do? He started to run. One hundred and fifty miles once again, up and down mountains, back to Athens. The return trip took him another two days; when it was all said and done, Pheidip- pides totaled three hundred miles in four days.

The Athenians couldn't wait the week for help. They didn't have the luxury of time and would have to find another way. The Athenian general, a man named Miltiades, ordered his troops to launch a surprise attack, a little more than twenty-five miles outside of Athens, in a town called Marathon. At dawn, the Athenians charged before the Persian troops had time to prepare. It worked, and Athens pulled off one of the most stunning victories in history.

As the legend goes, while the Athenians finished off the last of the Persians, Pheidippides noticed that a Persian ship that was sailing away started to change its course. It turned to point straight to Athens. Chatter spread throughout the Athenian ranks. Maybe the Persians planned to sail to Athens and proclaim victory. Not knowing any better, the citizens of Athens might simply surrender to the foreign army before their victorious warriors could make it home to tell the true story of their triumph over Persia.

A ship could sail fast, but Pheidippides knew he could run faster. So, he set off for Athens, running faster than ever. He never stopped once. He dropped his weapons so he could go more quickly, then tore off his armor so he could go quicker than that. Finally, he rushed into the Parthenon and announced, "*Nike!*" It meant, "Joy! We have won!"

Then, in his exhaustion, he promptly collapsed and died.

The latter part of that story is, according to historians, almost entirely false. Pheidippides did exist, and it seems as though he did actually run the three-hundred-mile round-trip to Sparta in a matter of days. But his death would have been more widely recorded if it did indeed happen on the steps of the Parthenon in the manner detailed in the legend. What happened was that as Greeks retold the story of the heroic Battle of Marathon, Pheidippides's role grew and grew. The truth got so twisted that, according to one telling, a Greek god visited Pheidippides on the road back from Sparta and prophesied an Athenian victory.

Pheidippides went from war hero to myth.

Why? Because myths are how we remember history, how we teach values, and how we connect to our culture. The story of Pheidippides, to the Greeks and to people today, exposes the virtues of patriotism. It's a foundational text for what it means to be Greek.

The Religion of Sports has its own myths, too.

///////

Sports *were* religion back in the days of Pheidippides. Once every four years, during the summer, a messenger ran from city to city across ancient Greece yelling, "*Ekecheiria!*" The word meant truce, and those cries literally stopped wars in the middle of battle, spears and shields dropping to men's sides. The soldiers rushed home, gathered belongings, and they and Greeks everywhere traveled to the most holy spot in their world, the base of Mount Olympus, to gather for five days of the Olympic Games.

The ancient Olympics didn't look like the sporting spectacles we know today. The Greeks created the games to honor Zeus, the king of the Greek gods, and they hosted them not in a stadium but in a temple. Greeks raced by foot and chariot, threw javelins and discuses, and wrestled, all in an effort to show Zeus that they had tried to model themselves in his vision, strong and physically gifted. Soon the Greeks modeled their gods after these athletes, too; they believed that the sun was actually Helios, a chariot racer whose path, every day, led him across the sky. Every night, having won his race, he celebrated by lounging across the heavens with his prize, a golden cup: the moon.

The Olympics continued in this way for nearly 1,200 years until a Roman emperor outlawed paganism. With the religious aspects banned, the Greeks abandoned the Olympics as well. But other cultures, worlds away, saw a connection between sports and the heavens, too. The Mayan ball game, the earliest team sport ever recorded, occurred in the shadows of temples, and the losing teams would be sacrificed to the gods. Even those in the Roman Empire saw the connections between sports and the spiritual. The apostle Paul, for his part, used sports to help him explain his own religion. "You know that in a race all the runners run but only one wins the prize, don't you?" he wrote. "You must run in such a way that you may be victorious. Everyone who enters an athletic contest practices self-control in everything. They do it to win a wreath that withers away, but we run to win a prize that never fades. That is the way I run, with a clear goal in mind. That is the way I fight, not like someone shadow boxing." (1 Corinthians 9:24–26).

In the 1880s, archaeologists uncovered the ruins of the original Olympic temples in Greece, and Europe flooded with stories of the ancient competitions, imaginations everywhere running wild. Perhaps nobody found more inspiration from those discoveries than Pierre de Coubertin, a Frenchman who found himself disillusioned after the French defeat in the Franco-Prussian War. He learned how the Greeks viewed their training as a religious act, how they considered keeping in shape to be as sacred an act as any. For years, Coubertin became obsessed with reviving the Games, with resurrecting the Olympic values. If he could bring back the Games, he thought, then maybe he could bring back that sense of pride in one's competitive spirit. Maybe he could inspire the type of person who could win France's next war. Finally, in 1896, he staged the first modern Olympics in Athens.

In his personal journal, Coubertin documented his journey to develop the Olympics. Over and over, he wrote of how he wanted to capture a feeling, a shared sense of values that he sensed had been present when the Greeks staged the Olympics. In other words, he didn't just want to create the Olympic Games; he wanted to create Olympism.

"The first essential characteristic of ancient and of modern Olympism alike is that of being a religion," Coubertin wrote. "The ideal of a religion of sport, the *religio athletae*, was very slow to penetrate the minds of competitors, and many of them still practice it only in an unconscious way. But they will come round to it little by little."

Come around they did.

When Coubertin planned those first Olympic Games, it was important to him that they were held in their ancestral home, Athens. That part was settled, but there was still the question of events: Competitors would box, throw javelins, and run sprints. But could there be something else? Coubertin wondered if there was some sort of signature event that they could stage, something that would connect these modern Games directly to the ancient Greeks?

Enter: the marathon. Coubertin and others concocted a long-distance race that would honor Pheidippides by tracing his fabled route: Marathon to Athens, 26.2 miles in total. The event was a smash hit. At the stadium where the race ended, one hundred thousand spectators waited for the runners to enter. A Frenchman led almost the entire race, but with about five miles left, he quit. Soon an unknown and poor Greek man had the lead, and when he ran into the stadium, the crowd cheered, "Greece! Greece!" Two Greek princes hopped onto the track and jogged alongside him as he finished the race, and when he crossed the finish line, the king told him that whatever the man could imagine, the king would grant him as a prize. The man asked for a donkey and a carriage. He sold water and said that the cart would lighten his load.

The legend of the marathon was alive again.

But something was wrong—the race only featured men. Couldn't women run 26.2 miles, too? For decades, officials barred them from the Olympic race and all the other marathons that popped up across the world. That all changed when a young woman named Kathrine Switzer laced up her shoes in Boston one day in 1967. For a recent project, the Religion of Sports team caught up with Switzer, and hearing her speak of the events that day, it was striking just how much her actions changed the world. Switzer's run is iconic, and the more she spoke, the more one thing became clear—hers isn't just a sports story.

It's a foundational myth.

///////

It's hard now to imagine the sentiment surrounding long-distance running and women in the 1950s and '60s. At the time, the idea that a woman could run 26.2 miles was laughable. In the early 1950s, many in the running establishment held on tightly to two beliefs: that man could never break the four-minute mile and a woman could never run longer than 800 meters.

Both, of course, seem ridiculous today. But that's the environment in which Switzer came of age.

Switzer remembers that as a kid, she ran around the neighborhood and heard people in their front yard tell her, "You shouldn't do that. You're going to turn into a boy." As she ran away, she'd hear their whispers. "What's going to happen to her?" they'd say. "You just don't know . . ."

Nobody knew because so few women even tried to run long distance at the time, at least not visibly. Some of the limitation came from a failed Olympic experiment. In 1928, women competed in Olympic track-and-field events for the first time. The farthest women's race was the 800 meters—half a mile, or two laps, around the track.

They raced on a warm Thursday in Amsterdam. Nine women made it to the final race, and they came out of the blocks blazing. They finished the race just about as fast, too: six of the nine runners beat the previous world record that day. The women rewrote the record book in real time.

But that's not how the press saw it. For whatever reason—perhaps the sportswriters were prejudiced or perhaps they didn't even attend the event and got poor secondhand information—they reported that five of the runners collapsed following the race. Video exists of the event, and if you watch it today, you'll see the runners cross the finish line, put their hands on their hips, and try to slow their breathing. One runner falls to the ground, briefly. It looks, basically, like the aftermath of any race.

But most people couldn't watch that video. Videos couldn't spread like they do today. Instead, people just read the papers, and the reports created a very different kind of myth, one that would permeate for decades to come. "It was not a very edifying spectacle to see a group of fine girls running themselves into a state of exhaustion," wrote the legendary Notre Dame football coach Knute Rockne, who worked those Olympics as a celebrity columnist. The *Times* of London said the race was "dangerous," and the *Daily Mail* claimed that by running farther than 200 meters, the women risked prematurely aging. In Montreal, the paper called for the International Olympic Committee to ban the event, saying of the 800 meters, "It is obviously beyond women's powers of endurance, and can only be injurious to them."

The IOC met following the race and, according to the *Los Angeles Times*, engaged in "long, hot debate." Officials emerged with an edict: the women's 800-meter race would be eliminated. The longest event that women could compete in would be the 200-meter. It stayed that way until 1960.

In that void came rumors. If a woman ran a mile, she'd get hairy legs. If she ran a 5K, she'd become infertile. If she ran a 10K, her uterus could fall out. And if she tried a marathon? Well, she'd probably die trying to finish. But if, somehow, some way, a woman *was* able to run 26.2 miles? Of course, she'd turn into a man by the time she made it to the finish line.

These are all the stories that Kathrine Switzer heard when she ran as a young girl.

These are all the stories she would prove false—once and for all.

/////////

Switzer ran because she had always run. Since she was a young kid, running has made her feel free. She never thought twice about what everybody said about her. Turning into a man? All of it was obviously ridiculous. Switzer *knew* it wasn't true because she ran every day, and last she checked, she was still a woman.

So, when she enrolled at Syracuse University to study journalism, it only made sense that she would keep running there, too. Even though there were twenty-five varsity men's sports, all with scholarships, women didn't have a single intercollegiate sport at Syracuse in the mid-1960s. But Switzer was competitive. She wanted to get better. So one day, she went to the men's cross-country team's practice and asked if she could run with them. They didn't believe in the pervasive myths, either. This was a college campus in the 1960s, after all; they weren't keen on accepting long-held beliefs. Sure, they told Switzer. Try to keep up.

She couldn't at first. She had never followed a traditional training regimen before. But as she practiced, day after day, something changed within her—and not just the fact that her mile times kept dropping. "I remember quite distinctly," Switzer said. "Once I got serious and ran over three miles a day, I stopped going to church. I realized it was because I felt closer to God and the universe out in nature than I ever did inside with a group of people."

A volunteer coach, named Arnie Briggs, took Switzer under his wing. Arnie was the fifty-year-old campus mailman, but running was his passion. He had finished the famed Boston Marathon fifteen times, and while Switzer trained with him, Arnie told her about the starting line at Hopkinton High and Heartbreak Hill. He told her story after story. Switzer loved hearing them, and she wanted more. "Let's quit talking about the Boston Marathon and run the damn thing," she said one day.

Arnie, who ran with Switzer every day, said no. "No woman can run the Boston Marathon," he told her. No woman had ever run the Boston Marathon officially, registered through the race with a bib and everything. Even Arnie, it seemed, believed at least some of those old myths.

Switzer didn't. She never believed what everybody told her about female athletes. She pushed back, and finally, Arnie relented. "If any woman could do it, you could," he said. "But you have to prove it to me." Switzer would have to finish the distance in practice first. If she could do it, then Arnie would personally take her to Boston—where she could become the first woman to ever officially run the Boston Marathon.

They ran the practice round three weeks before the real thing. The miles passed, and Switzer felt good. Arnie agreed. "You look strong," he told her. She felt strong, too. She felt so strong, in fact, that when they reached the end of the route, Switzer didn't believe they had gone the full distance. "Let's go five more miles," she told her coach, and when they finished the thirty-first mile, Arnie passed out.

When he came to, he looked up at his protégé with something like a wink in his eye. "Women," he told Switzer, "have hidden potential in endurance and stamina."

The two of them went through the Boston Marathon's rule book and found that at no point did it say that women could not compete. It seemed as if the organizers just assumed that women wouldn't even try. So, with Arnie looking on, Switzer filled out the registration form. When it asked for her name, she signed it "K.V. Switzer." She devoured the works of J. D. Salinger in those days, and thought it was neat to go by her initials. Never did she think that it would appear as if she was concealing her gender to race officials.

The entry fee was three dollars, and Switzer pulled out three one-dollar bills. Arnie put it all in an envelope, and being the mailman and all, sent the registration materials on their way to Boston. Switzer would be on her way soon enough.

///////

Unlike a lot of the marathons today, the race in Boston in 1967 didn't start until noon—so Switzer and a group of friends gorged themselves at breakfast. Switzer ate bacon and eggs, some pancakes, extra toast, and a cup each of juice, coffee, and milk. She needed all the fuel she could get.

The race was held on April 19, Patriots' Day, which is a holiday in Massachusetts. No, Patriots' Day isn't a day of recognition for the football team from Foxborough (we call that annual holiday the Super Bowl victory parade); Patriots' Day commemorates the Battles of Lexington and Concord, which started the American Revolution in 1775. Since 1969, the day has been observed on the third Monday of April, but when Switzer ran, it was observed on the actual date. In 1967, that meant race day was a Wednesday. The day isn't a holiday in New York State, so Switzer would have to hurry back to Syracuse for classes the next morning.

In Massachusetts, though, all the local schools are let out, the Red Sox play a day game, and everybody in town watches the Boston Marathon, which started in 1897, one year after Coubertin's resurrected marathon inspired similar events across the world. No other marathon existed in America—and to this day, Boston remains the oldest annual marathon in the world. Early race organizers wanted to connect the American Revolution with the Greek democracy's own battles for independence, to recast Paul Revere in the image of Pheidippides. In America, we retell our myths through sports.

One year, when I was young, Mom took my sister and me to the finish line. We cheered when racers passed, and it struck me that unlike basketball, where everyone was tall, or football, where everyone was big, there was no one type of runner. The young and old ran next to each other, and they were White, Black, Brown, and everything in between. My sister and I cheered every one of them, and they looked at us and waved and got an extra pep in their step. They were the runners, but we had an important job, too. Our support gave them strength, and we were all a part of the race's process—each in our own way. That's one of the things that is so special about the Boston Marathon: it's a celebration for everybody, a sprawling city linked by sports. It's thanks to Switzer that women are included in that community, too.

///////

Often, Patriots' Day is the most gloriously sunny day of the year, as much a symbol of spring as Easter, Passover, or Holi. But in 1967, freezing rain fell all along the course, and wind howled through the streets. The Red Sox game got rained out. Switzer put on gloves, a gray sweatsuit, and gold hoop earrings. She pinned her bib—number 261—to her shirt and headed to the starting line.

"I have often said that a run every day is like praying," Switzer told the Religion of Sports team one spring day approaching the anniversary of her famous run. She lives in New Zealand now part-time, so that when it's winter in America, she has perfect weather to keep running. As she spoke to us over Zoom, she stopped once to admire a bird outside. She spun the camera around to show us and revealed not only the bird but also a sloping, green hillside that looked straight out of a travel magazine. Not a bad view for a jog. She was recovering from a mild back injury but still ran every single day. To Switzer, those daily runs are different than races. She remembered what it was like near the starting line that day in 1967. "When we go to the marathon," she said, "it's like we've gone to Mecca. It's like a pilgrimage. Once in the marathon itself, we're there together. We're pilgrims trying to cover this distance together for whatever cause we believe in."

Switzer tried to keep her head down in the starting area, but some fellow pilgrims noticed she was a woman and, to her surprise, they welcomed her warmly. These were not the sexist groups who believed the myths about women runners. They gave Switzer encouragement and asked her to give their wives tips to get into running as well. This was her tribe. A race official looked at Switzer's bib number and moved her forward. She was in. Now she just had to run.

//////

That's exactly what she did. She ran one mile, then two, and slowly, her nerves started to fade. She might have been on the big stage of Boston, but this was still running, after all. She found comfort in the familiarity of the action. Nothing she did now was much different than what she did as a kid running through the neighborhood. That's another thing that's distinct about sports—even as the stakes get higher, the fundamental heart of a sport is the same one that's played at recess and in backyards everywhere around the world. Each step calmed Switzer's nerves and made her realize that she actually could do it; she could finish the race and make history.

But then, at mile four, came something that had never happened before during Switzer's runs. A flatbed truck barreled through the course. It was the press. They had just learned that Switzer was trying to become the first woman to officially run the race, and they wanted to know why. Flashbulbs exploded. Journalists hollered questions like, "When are you going to quit?"

Amid the ruckus, Switzer heard someone running after her. She swung her head around to see who it was. A man in a suit was coming after her, and he was angry. "Get the hell out of my race and give me those numbers!" he screamed. The man tried to grab Switzer's bib number on the front, but he missed and tried to grab the one in the back. Arnie tried to push the man away, and Switzer just tried to keep running. The cameras on the truck snapped photos. A few of them would become some of the most famous photographs in sports history. The man reared to grab Switzer's bib numbers again, Switzer started to brace for impact, and—suddenly, Switzer's boyfriend came from behind to bodycheck the man. The man splayed out in the middle of the road and Arnie yelled, "Run like hell!" So, they did.

The man was Jock Semple, and he was the director of the Boston Marathon. He hadn't realized that a woman had registered for his race, and he wasn't going to have it.

Switzer wouldn't have it, either.

///////

This is the part of the story where Switzer turns from an incredible athlete into something far greater. It's where she becomes mythic. You might have heard of Joseph Campbell and the hero's journey. Campbell was an academic who studied myth, and he believed that stories of the ancient Egyptians, the Bible, and *Star Wars* all followed similar patterns. They taught lessons and traced a path that he dubbed the hero's journey. Through the retelling of those stories, we provide the vessel through which many of religion's—or life's—lessons can be taught. Myths, Campbell once wrote, "could even show us, as seen from below, how the soul views itself." I'm constantly leaning on his theories when I'm telling my own stories about athletes. There's something visceral about how the myths of sports capture the essential truths of life.

Here's a traditional myth, told in the most bare-bones version possible: Achilles becomes a great hero in *The Iliad*, helping the Greeks defeat the Trojans. He's the great warrior, destined from birth for glory. After some ups and downs, he fulfills his prophecy.

It's a great story and all, but it feels distant. The Trojan War? Isn't that when USC plays UCLA? We don't know what Achilles looked like exactly. We only heard of his feats secondhand. With sports, we actually live our myths.

Here's a similar story, but told through sports: LeBron James becomes the great warrior, destined from birth to be, as *Sports Illustrated* declared while he was in high school, "The Chosen One." LeBron joined his long-suffering hometown team, couldn't quite win, left, gained some experience, and then returned to finally make the Cleveland Cavaliers champions.

Which one feels more impactful? LeBron, right? His story is more applicable, believable, *real*. And of course, that's because it *is* real. We all watched LeBron win that championship with superhuman strength. We saw the impact he had on his hometown. There was a giant Nike poster in downtown Cleveland at the time that showed LeBron, arms outstretched in black and white. Above him were the words WE ARE ALL WITNESSES. Sports are like having a front row seat to the Greek gods.

Few people I have met have lived quite as mythic a life as the Paralympic track star Scout Bassett. When I first met Scout, one of the first things I noticed was her height. She's tiny, just 4-foot-9 and 85 pounds. She wears a size 11 shoe—in children's size. But she has a presence about her, a steadfast confidence, that fills any room. As she started telling her story, there weren't many dry eyes among our crew.

Bassett was born in Nanjing, China, three hours inland of Shanghai. At only eighteen months old, she was involved in a chemical fire that damaged her right leg. Her parents abandoned her on the side of the road. Someone found her and took her to a police station; police took her to a government-run orphanage.

They amputated her leg, but the doctor performing the procedure did it in haste. Years later, a bone grew out of the back of Bassett's thigh. Can you imagine the pain? There wasn't enough money for a proper prosthetic, so she fashioned a homemade leg, connected to her body with leather straps. Bassett told us that she barely went outside for six years. "I endured unimaginable pain and loss and trauma," she said.

When she was eight, an American family adopted her. She'd never heard of America, never seen a white person, didn't speak a word of English, and weighed only twenty-two pounds when she boarded a plane for America. And there it is—our hero's call to adventure.

The transition wasn't easy for Scout. She moved to a small town on the coast of Lake Michigan with about 1,100 residents, most of whom were White. She was the only Chinese student in the school—and on top of all that, she had only one leg. Scout got a proper prosthetic, but always wore pants. She thought that by doing so, people might not realize that she was different.

At fourteen, her adoptive mother took her to Orlando, Florida, to meet with a renowned prosthetist named Stan Patterson. He wanted to make Scout a running blade. "You're going to be an athlete," he told her.

Every hero needs a guide on their mythic journey.

Scout balked. She was already signed up for the volleyball and basketball teams, but they never let her leave the bench. She was only 4'6" at the time. What kind of athlete could she be?

Patterson had an idea. Bassett had come to visit at exactly the right time. Disney's Wide World of Sports Complex, just down the road, was hosting a track meet for disabled athletes the very next day. He knew the organizers. He signed her up for the 60-meter race.

Scout was terrified, and when her mom drove her to the meet the next morning, Scout later described, "I start having the meltdown of all meltdowns." She cried. She screamed. Most of all, Scout stared at her new running blade. She was in shorts, and everyone could see that she had a disability. "That was the first time in my life that I was going to have to step out onto a track, be in a lane by myself, and really show the world who I am unmasked," she said. "I just didn't know if I was ready to do that."

Scout made it to the starting line, and when the starting gun went off, she ran forward. Sixty meters isn't very far, but unlike Kathrine Switzer, Scout hadn't spent her whole life running. It was the farthest she'd ever gone. And what happened? As she remembered, "I came in dead last by a lot, but it was just this feeling of the chains that held me down as a young girl, the struggles that I had with identity, of self-confidence, self-belief, self-esteem—all of that was lifted.

"I said to myself, 'I'm never going to be ashamed of what I look like, where I come from, of my story, and of the things I cannot change about myself.'"

In Campbell's terms, the moment signifies the beginning of Scout's transformation, the crossing of a threshold.

Scout dedicated herself to training, went to UCLA on a full-ride scholarship. But after graduation, she quit her job to train full-time. She slept in her car, and it eventually paid off. "I went from totally not in the picture to top five in the world rankings," she said. Best of all, she gained a sponsor, Nike, and was able to move into her own place.

Then Scout ventured off, being a hero. The Paralympics in Rio. Motivational speeches. Nike approached her and asked if she'd tour China to help them launch a program about youth sports. Scout agreed—but asked if she could visit the orphanage where she grew up during the trip. When she went back, Scout walked the halls stoically. She played with kids. She found her old room. "Up until then, I did not really realize that I had a lot of harbored pain from my childhood," she says. "I remember that the one thing I never felt as an orphan was a sense of hope and a sense of love. When I see those kids, I wanted them to know that they matter, that they're important, that their life matters."

When she left, Scout cried. "For days," she says. "For weeks. For months."

Scout felt numb. Even running and training felt meaningless. "I just went down the darkest of tunnels that I did not know if I was going to be able to recover from," she said.

Before triumphing, every hero has to hit the abyss—to metaphorically be reborn.

Little by little, Scout moved forward. "I told myself to lean in to everything that happened, lean in to the work that you've done and use that as a source of strength and power," she said.

Her next big race came in London, at the 2017 World Para Athletics Championships. Scout had qualified for the 100-meter race, and as they introduced the competitors, Scout sobbed while walking around the track and up to the starting blocks. All she had been through, all the trials: it all came to a head in this race.

Then it was time to run.

Scout was by far the shortest competitor in the race; some of the athletes she ran against were more than a foot and a half taller than her, which makes a massive difference in a sport like sprinting, where stride length is of the utmost importance. Scout has to work twice as hard just to go the same distance. Here's how ESPN once described Scout's height: "It's like a 150-pound man playing offensive tackle in the NFL."

Scout got off to "a terrible start." At 50 meters, she was in fifth and far from medal contention. "I knew I had what it took," Scout said. "But I had a lot of work to do in those last fifty meters."

She ran. Step after step after step. Soon, she passed one competitor. Then another. At the finish line, she strained forward. She finished third. She would stand on the medal stand after all.

"Everybody always thinks about the record-breaking moments or the gold medal moments, but for me, winning that bronze medal is one of the things that I'm the most proud of," she said.

As for what got her to that point? Scout says she wouldn't change anything—no matter how challenging her road to that race might have been. "You've got to be willing to face the very things that per-

haps have broken you, to be able to slay those dragons and those demons and to get on that podium," she says. "That, to me, is greatness."

Indeed. Scout slayed the dragon. Her journey is complete.

We need these types of myths because they inspire us. They give
us role models to pattern our behavior after. They help us understand
the world. Aristotle once wrote, "A lover of myth is in a sense a lover
of wisdom." By that logic, fans have learned the wisdom of the world
through sports. There is so much that anyone can take from Scout's
story, such as resilience, belief, and that we should never judge a book
by its cover. Jack Nicklaus winning the 1986 Masters showed that
age is just a number. The 1980 U.S. Olympic hockey victory over the
Soviet Union made us believe in miracles. The New Orleans Saints'
Super Bowl victory in 2010 helped heal a city reeling from Hurricane
Katrina. There's something intensely spiritual about all these moments. "Nobility of spirit," Campbell explained in *The Masks of God*,
"is the grace—or ability—to play."

Kobe Bryant adored Joseph Campbell. He read *The Hero with a
Thousand Faces* cover to cover, and when we were together making
the documentary *Muse*, we would have long conversations about
Campbell's theories. "How do we take sports and tell beautiful tales,
beautiful stories that connect to human nature?" Kobe once said. I
think that the answer is that sports *already* tell beautiful stories that
connect to human nature. In a way that no other aspect of our lives
can, sports provide a canvas in which greatness can happen, where
humans reach and surpass their limits, where anyone can become a
hero—but more on Kobe later.

We have a race to finish in Boston.

///////

When the race official attacked Switzer, something incredible happened. She was intimidated, of course, and she was scared. But she

didn't quit. Talking to her today, it's striking how mature her response was back then, even though she was only twenty years old on race day. After running a safe distance away, Switzer turned to Arnie. "I'm going to finish this race on my hands and my knees if I have to," she said. Why the sudden determination? "I had to," she explains, "because women were always being told they were barging into places where they're not welcome, and they can't do it. I *knew* I could do it. And I knew—as embarrassed and scared as I was—if I dropped out, everybody would say, 'See? She's just here for a joke.'"

Something else happened as she ran. The farther and farther she jogged, the clearer her head became. Through the action of running, she found something like clarity. The anger she felt toward the official faded. The anxiety that she, a young college student, had about the future vanished, too. She found her purpose. "It's my job now to figure out how to get women the opportunity so they are not afraid, and they will do something bold and experience this wonderful feeling I have of freedom and destiny and transcendence," Switzer told us of her thinking during the race. "And I really wanted to give them that. But how, though? How?"

The specifics would come later. What she realized is that she didn't really have much of a choice whether she would become a symbol or not. With the flash of a camera, Switzer transformed from a woman trying to push her limits into a global icon. When she was attacked, she stopped running for herself and started running for women everywhere. It was her call to action. She couldn't stay anonymous if she tried. But the attention didn't overwhelm her. Quite the opposite, actually. The more weight got added to her shoulders, the more steadfast she became. That's an extraordinary load to carry as a twenty-year-old, or at any age for that matter, but that's the thing with mythical heroes. They accomplish the extraordinary.

Switzer ran with that weight all the way to the finish line, where the officials registered her time. She did it: Kathrine Switzer was the first woman to ever officially complete the Boston Marathon.

"My dad had always taught me when you do something, you've got to finish it and take responsibility for it," Switzer told us. "I always firmly believed that, so I picked it up and have been responsible for it for the rest of my life."

The next morning, when she walked to class at Syracuse, she passed an area where the journalism school displayed newspapers from all over the world. She looked up. Her photograph was plastered onto the front pages from New York to London to Tokyo, #261 escaping the hotheaded official.

/////////

Switzer still runs nearly every day, even at seventy-four years old. She has run forty-two marathons since 1967. In 1972, she ran Boston again, with a whole group of women this time. She had fought the same official who attacked her to make sure that women would be allowed in the race, and after years of lobbying, she succeeded. She ran Boston in 2017 also, the fiftieth anniversary of her run, where she became the first woman to ever run marathons fifty years apart. She clocked in only twenty minutes slower than her twenty-year-old self—and that's while being hounded for photos the entire length of the race. "That's not because I'm great," she says. "It's because of how few women ran fifty years ago."

Switzer has made sure that fifty years from now, there will be millions more women like her. That figure is no exaggeration. After graduating from Syracuse, Switzer taught herself the ins and outs of business, about sponsorship, marketing, and how to craft a proper proposal. She sent a pitch to Avon Cosmetics, the largest cosmetic company in the world at the time, about how they should sponsor a series of races specifically for women. There was a huge untapped market, she said. "We'll never do anything with running," the people at Avon told Switzer. "But we'd love for you to work for us."

Switzer was fresh out of college. A full-time job? "That's good enough for me," she told them. But she kept hounding her managers about the race. Just do one event, she begged. Finally, they did. It was a success. It was *such* a success that they adopted the entirety of Switzer's initial proposal, hosting four hundred races across twenty-seven countries. More than a million women ran.

Then Switzer turned her attention to the IOC. It was 1976, and the Olympics were coming stateside to Los Angeles in 1984. She wanted to make sure that when people all over the world tuned in, they saw women running the marathon. Typically, it takes decades to get new events approved for the Olympics. Switzer says that if she had followed typical IOC protocol, women wouldn't have run in the marathon until 2012. But she fought for equality with haste—and on August 5, 1984, fifty women from twenty-eight nations raced through the streets of Los Angeles for 26.2 miles in the first women's marathon in Olympic history.

"Two point two billion people watched it on TV, and everybody knows how far 42.2 kilometers, or 26.2 miles, is," Switzer says. "It changed a lot of notions. In fact, that was as important as giving women the right to vote, because it was the physical equivalent of that intellectual, social acceptance in 1920."

According to a survey of international marathons, females made up 32 percent of all competitors who completed a race in 2019. That figure grows about 2 percent each year. And you can bet that if you watch runners pass by on race day, you'll see little pieces of Switzer with them somewhere. Her bib number, 261, has become a sacred good-luck charm for women runners everywhere. Switzer reads the emails from strangers. "I'm wearing number 10,942 in my first New York City Marathon tomorrow," one woman said, "but I wanted you to see what I was wearing on my back." Pinned to her back is a bib with a different number: 261. Runners take a Sharpie and write the three digits on their arm, wanting to carry some of Switzer with them as they cross the finish line. Some of them go to tattoo parlors after the race and get the number inked permanently.

"I had to think that if somebody is tattooing 261, what does it mean to them?" Switzer says. "They kept using the word *fearless*. I suddenly understood that they were relating to the story of being told that they weren't good enough, or that they didn't belong, or that they were the wrong color or the wrong religion or nationality. They went out and ran anyway. And then, they felt fearless. They said, 'Well, I can do this.'"

Switzer summed up her work succinctly: "Through the vehicle of running, we get women to feel fearless."

There's a story Switzer told of the moment she finished the race in Boston on that cold day in 1967. The press waited for her near the finish line; the same journalists who hounded her on the back of the press truck had been waiting for her there. They greeted Switzer about as warmly as the freezing rain that continued falling. One reporter said, "This is just a one-off deal. You'll never run another marathon again, right?"

Switzer stared back blankly. "You know what? One day you're going to read about a little old lady who's eighty years old and drops dead on a training run in Central Park," she said.

"It's going to be me."

Switzer, it seems, was getting herself confused with Pheidippides. But unlike his, we know for a fact that her myth is real. It's part of the Religion of Sports. There's no melodramatic moment of dropping dead. It's captured forever in that famous photograph: Switzer just running and running and running, not letting anything stop her, giving women strength with each stride.

KATHRINE SWITZER

NICKNAME: The Marathon Woman

DESCRIPTION: First woman to officially finish the Boston Marathon, activist

FACT #1:

Shattered barriers in 1967 when she became the first woman to finish the Boston Marathon despite being attacked by race officials.

FACT #2:

Successfully lobbied the International Olympic Committee to include the women's marathon in the 1984 Olympics, which was viewed by hundreds of millions across the world.

WISDOM:

"When I go to the Boston Marathon now, I have wet shoulders—women fall into my arms crying. They're weeping for joy because running has changed their lives. They feel they can do anything." —*Kathrine Switzer*

CHAPTER 4

Transcendence

Get in the zone and tap into a higher power.

I always turn to the sports pages first, which records people's accomplishments. The front page has nothing but man's failures.

—Earl Warren,
late chief justice of the U.S. Supreme Court

BODHIDHARMA COULD HAVE been a king. He was the prince of a region of India, born around 500 AD, and the youngest of three sons. He was the king's favorite, and the whole kingdom could have been his. When the king died, Bodhidharma sat in front of his father's coffin for seven days, meditating. What would he do next? Would he become king after all? A week later, he found some clarity. Bodhidharma decided he had no interest in politics. He didn't want the throne. He yearned for something deeper, something with more meaning. He left the palace and traveled to a monastery, where he studied to become a Buddhist monk under a famous master.

The teacher trained Bodhidharma, and for weeks on end, the two of them would meditate, discussing subtleties of the technique.

"Master," Bodhidharma asked once, "when you pass away, where should I go? What should I do?" The teacher instructed his pupil to travel to China, where he would find an audience for his teachings.

For the next several years, when his master was still alive, Bodhidharma taught throughout India and Persia, preaching that wisdom can be realized through intense meditation. His following grew and grew. Among those who sought out his wisdom was his brother, who had decided to take the throne and was now the king. Bodhidharma became one of the most respected and powerful religious figures in India, but when his master died, he decided to listen to his teacher's guidance. Soon rumors flew all over the kingdom. *Bodhidharma is leaving! He is going to China!*

The king tried to convince his brother to stay, but nothing could change Bodhidharma's mind. They were still family, though, and the king sent a message ahead of him. *Take care of the traveling monk who will soon arrive in your country.*

In China, anticipation grew and grew. *What kind of monk was so important as to merit a message from a king?* When Bodhidharma arrived, a large, raucous crowd had gathered to hear him speak.

Bodhidharma sat in front of the group and began to meditate. He remained entranced for several hours. He said nothing. The audience laughed. Some of them tried to follow along, then gave up. Others grew more and more angry. Eventually, Bodhidharma just stood up and walked away. The crowd didn't understand, and so Bodhidharma continued with his travels, searching for the followers that his master predicted would be there.

After the incident with the group, Bodhidharma became infamous throughout the region. This Indian monk, people whispered, sure had a lot of nerve. The emperor was intrigued and called Bodhidharma to the palace. He wanted to see what the fuss was about. Bodhidharma arrived, and again he meditated. The emperor was furious at the apparent lack of respect. He banished Bodhidharma from his kingdom; once more, Bodhidharma's message had fallen short.

Exiled, the monk kept traveling. He journeyed north until, finally, he came upon a temple surrounded by mountains known as Shaolin. The monks at Shaolin had heard Bodhidharma was approaching, and they came outside to greet him. They invited him to stay in the eastern wing of the temple, but Bodhidharma didn't even stop to say hello. He didn't say . . . anything. He felt called elsewhere, and so he climbed up past the temple into a mountain cave. He went inside, sat down, and, once again, he started to meditate. He was searching for something, for some kind of wisdom. Other monks would hike from the temple to visit him, but he would never respond. They'd ask him to train them, and he wouldn't acknowledge their pleas. He was too focused on his meditation. One time, the legend goes, Bodhidharma lost his concentration and started to drift off to sleep. So frustrated at himself for his lack of discipline, Bodhidharma cut off his eyelids so the lapse could never happen again. In all, he stayed in the cave for nine years—long enough for the dark outlines of his shadow to be imprinted onto the cave's wall. You can still go visit it to this day.

During his meditation, Bodhidharma had come to discover some essential truths about life. Today he's considered to be the founder of Zen Buddhism, and when he returned from the cave to the temple, he was welcomed again by the monks at Shaolin. This time, he returned their greetings. And now, finally, he agreed to teach them. He wanted them to gain the same insights that nine years of meditation had unlocked for him. But how? How could people feel that same type of inner peace? Whenever he had tried to teach his philosophy to others in China before, he had failed.

Legend says that Bodhidharma started to lead the monks at Shaolin through a new type of movement. To train their minds, he decided to train their bodies. It wasn't feasible to have them all meditate in a cave for nearly a decade, nor could he make them feel the need to cut off their own eyelids—so he led them through a series of drills and exercises. He based the movements on eighteen different animals, including snakes, tigers, and leopards. The more the monks practiced, the more they lost themselves in the movement. Time slowed. The chatter in their brains stopped. They started to understand just what lessons, exactly, Bodhidharma was trying to impart. They kicked and punched over and over again, following their master.

Bodhidharma, the story goes, came down from his cave wanting to teach others how to feel the same clarity that he did. And that is how, over two thousand years ago, Bodhidharma came to invent the sport that we now call kung fu.

//////

What Bodhidharma felt while meditating in that cave, what he tried to pass along through kung fu, what countless others strive for in their religious or spiritual practice is to reach a mental state known as "flow." Outside of a religious context, tapping into flow state has become a trendy buzzword recently, especially among business executives. The *Harvard Business Review* is flooded with articles detailing best practices for cultivating a company culture that helps its employees reach a flow state. The U.S. Air Force commissioned research into the effect that flow has on its pilots, and they discovered that flow can lead to "a two-fold improvement in how long a person can maintain performance." A TED Talk about flow has garnered more than 6.5 million views and claims that losing yourself in the feeling is "the secret to happiness."

But what, exactly, is flow? According to psychologist Mihaly Csikszentmihalyi, who first identified the phenomenon, flow state is the feeling that one gets when one is fully present and immersed in a task at hand. He describes it as "a state of concentration so focused that it amounts to absolute absorption in an activity . . . people typically feel strong, alert, in effortless control, unselfconscious, and at the peak of their abilities. Both a sense of time and emotional problems seem to disappear, and there is an exhilarating feeling of transcendence."

Flow is the incredible thing that happens when you feel like you have divine inspiration, when all your worries melt away, and you are utterly enveloped by whatever it is that you're doing. It's freeing and exhilarating.

The existing body of research around the topic is only a few decades old, but already, what researchers have discovered is nothing short of extraordinary. Studies have found that flow is associated with an increase in happiness, productivity, and intrinsic motivation. Scientists have performed brain scans and found that reaching a flow state causes your brain to stop acting consciously and forces it, instead, to act in overdrive. The prefrontal cortex, which functions as your brain's critic and causes doubt and second-guessing, shuts off. Your brain can typically process 60 bits of information a second. In flow, that figure increases to 120. A hefty cocktail of hormones, like endorphins, dopamine, and serotonin, that leave us feeling happy and ready to perform, flood into the brain. Another study found that being in flow actually causes people's pulses to slow and their breathing to deepen. Researchers have discovered that people in flow activate the same muscles that are used to smile, literally forcing someone to feel happier. In other words, the psychological experience of flow can literally change a body's biology.

Letting go of worries, sensing a call to action, and feeling connected to a larger power are all the types of feelings that typically are espoused by religions. So, it comes as no surprise that nearly every religion utilizes flow in one way or another: Buddhists, like Bodhidharma, meditate. Muslims dedicate time every day to prayer. Members of the Navajo Nation chant, with drums and rattles, to create something like a trance. Every religion has its own tools to help its followers tap into that elevated feeling we know as flow.

Sports has it, too. It's called being in the zone.

//////

When you think of Golden State Warriors forward Draymond Green, you think of flying elbows, groin shots, and barbaric screams. You do *not* think of meditation, reiki, chakras, or flow state. Yet in the weeks leading up to the 2021–2022 NBA season, my team persuaded Draymond to sit across from my father, Deepak, to learn about meditation and other New Age techniques. The project became the Prime Video special *The Sessions*.

Draymond and my father seemed like an odd couple at first, but Draymond was open to learning—and my dad embraced the dialogue as well. They sat across from each other on a beautiful day when my dad started to explain things as only he can. "You're a basketball player, so you know there are moments when you are playing when you lose all sense of self," my father said. "Have you experienced that?"

"Absolutely," said Draymond.

"Have you experienced losing all sense of time?"

"Absolutely."

"That's the secret," my dad said. "Inner stillness and outer dynamic activity at the same time—this is what the great spiritual traditions call 'flow.' You've heard the expression 'to be in the zone,' right?"

"Absolutely," said Draymond.

"That phrase comes from 'to be Zen-like,'" Papa said. "And Zen is actually a derivative of the word *chán*, which means meditation. So meditation is not *how* to change your mind, but how to go *beyond* the mind to that place of stillness, where it doesn't matter what's happening around you: chaos, confusion, the crowd is screaming—you're totally independent of that. You're right there, in the moment, in the flow."

Draymond understood the connection, and I bet you do, too. You've probably heard athletes talk about how they've had moments when the game slowed down for them, and everything seemed to open up. You've maybe seen a gymnast perform an impossible floor routine, only to make the whole thing look effortless. Maybe you're a cyclist, and on the fifteenth mile of your Sunday morning ride, you realized that you hadn't really been pedaling for the last twenty minutes—at least not consciously. Something automatic occurred instead, and you lost yourself in the rhythm of your bike, miles slipping away.

That's the zone. That's flow.

So much about sports is centered on maximizing achievement, on reaching what is known as "peak performance." How fast can we run? How high can we jump? How many goals can we score? But peak performance might actually just be another name for a flow state. Sports, it might be said, are a constant journey to reach a state of flow—not unlike the years Bodhidharma spent mastering meditation. The results of that quest can be the moments that convert the rest of us into true believers.

"Flow," says Susan Jackson, an Australian researcher who is one of the world's leading sports psychologists, "sits at the heart of the majority, if not all, of the greatest athletic performances." Michael Sachs, a professor at Temple University and another prominent researcher in the field, agrees. "Every gold medal or world championship that has ever been won, most likely, we now know, there is a flow state behind the victory," he says.

For a long time, I considered being in the zone as something fleeting, something that was simply the result of a player being particularly locked in at the moment. But as I have spent more and more time with elite athletes, I've come to view the zone as a spiritual exercise more than anything else. It is a feeling that every athlete—*every single one*, from Little Leaguers to Major League all-stars—has experienced at one time or another. It's the feeling that sports, maybe better than anything else, helps us tap into again and again.

///////

"Faith," Steph Curry once told me, "is believing in the unseen and believing in a higher power."

Steph is, in a traditional sense, a devoutly religious man. He's a Christian. He grew up going to Central Church of God in Charlotte, North Carolina, every Sunday as a kid. His mom, Sonya, would wake Steph and his siblings at 6 a.m. so that they could read devotionals before the school day. If you ask him to describe himself, the first word he'll use is *believer*.

Steph is also, by any measure, the greatest shooter of all time. He's hit more three-pointers than anybody to ever play professional basketball. His ability to drain a three from anywhere on the court literally reinvented how the game is played. Steph's father, Dell, played in the NBA, but shot making isn't just baked into his DNA. Steph worked on it. As a kid traveling with his dad, he'd compete against NBA stars in H-O-R-S-E and often win. Steph was a late bloomer, so he had cockamamie shooting mechanics to compensate for his height. When he finally sprouted up, he and his dad worked for hours on their home hoop, totally transforming Steph's shot. To this day, he'll shoot a thousand shots before practice. He told me that if he's ever anxious, if he needs to find a little clarity before making a major decision, he'll find a court and shoot some threes. If you were to total up all the time that Steph Curry has spent working on his shot, you'd pretty quickly blow past the hallowed ten-thousand-hour mark.

One day in 2019, while filming for a documentary, I asked him about his shot. I just wanted him to dissect it for me, to describe in detail everything that went into one of those famous threes. "It starts with your last two steps," he said. "I don't look down at the three-point line. It's all just a feel of where you are on the floor. Everything is in motion, flow, and rhythm."

He continued, "I've worked on my release so many times that all the great moments I've had training combine into the game situation where you just let it ride and trust it that all that preparation and work that I put into it will lead to the ball going into the basket. It's all about the faith I have in what I do and living with the result of that. When I let it go, I have ultimate confidence and faith that it's going to work out in my favor."

When Steph shoots, he doesn't think. He doesn't have to, at least not anymore. It's all muscle memory at this point, and his body starts to take over and work on its own. In this way, Steph is like a master meditator. At will, with just a ball and a hoop, he can summon a state of flow. He takes those last two steps, knows instinctively where the three-point line is, and, as he says, lets it ride. I thought back to his definition of faith: the belief in the unseen and a higher power. "Steph," I told him, "when you take a three-pointer, you're praying."

He pushed back. He prays in church, he said. He prays before meals, before games. He said that they were different.

"Is it really?" I asked. "Is it really that different?" If you really think about it, shooting a three-pointer is an act of faith. Everything leading up to it is a discipline, a repeatable ritual that ripples through a shooter's legs, arms, elbows, hands, wrists, and fingers. But once a shooter lets go, it's just a belief. An act of faith that it'll go in.

He thought for a moment. "I guess you're right," he said. He talked about how his mind feels at peace when he shoots, how he gains clarity, how he feels the presence of something greater than himself. Maybe shooting was a prayer after all. "I never really thought of it that way," he said.

///////

Whenever I spend time with an athlete, I always find it curious which moments they remember from their career. Once, I was asking Tom Brady about two-minute drills, Super Bowls, and touchdown throws. "Which play was your favorite?" I asked.

It wasn't a Super Bowl that he remembered. It wasn't a game-winning Hail Mary. It wasn't even a touchdown.

It was one throw, a ten-yard curl to Donté Stallworth in a midseason game on the road against the Bills in 2007, right in the middle of the Patriots' undefeated season. The play was about as unremarkable, on its surface, as it could get. It was certainly not the most important play of that year. I'm a die-hard Patriots fan. Even I could barely remember that game and certainly not that play.

Recalling that moment, Tom couldn't help but smile. I've never seen him so giddy. "I almost giggle every time I think about that game," he told me. "Because ever since that game, I'm still trying to get back to that point, because in so many ways it was a perfect night."

The specific throw came in the beginning of the third quarter, 12:25 left on the clock. The ball sat at midfield, and it was first and 10. New England already led 35–7. Again—in every conceivable way, this was one of the most boring parts of the season. This was the part of the game where you feel like you can start flipping over to other channels to see what's happening around the rest of the league. The Patriots' offensive coordinator, Josh McDaniels, called a deep curl to Stallworth. Brady dropped back, waiting for Stallworth to finish his route. He could only see the back of his receiver's #18 jersey when Brady noticed the safety start to hesitate. Automatically, his instincts took over. He planted his feet, reached back, and fired a bullet down the field. "You feel like at that point, you can't do wrong," Tom described. When Tom released the ball, Stallworth had just planted to turn around at the end of the curl. And so, when Stallworth turned around, before he even had time to realize what was happening, the ball zipped in and hit him in the chest. All the receiver could do was try not to drop it, and he didn't. For Brady, that play with its throw, accuracy, and anticipation is the pinnacle. "That's a masterpiece in football," he said. "That was a clinic. That's the thirst that can never be quenched. That's the one you're always trying to strive for."

We moved on to other topics, talking about that season more generally. But Tom couldn't let that play go. "We did something," he said, "that I'm still trying to repeat today."

But why? What was it about that play that lent itself to being some kind of North Star? It had to be something deeper than just being performance based. I pushed Tom to explain what it was that he was chasing. Tom thought for a moment. He was trying to figure out the best way to verbalize it in a way that someone who didn't play at his level could understand. "The best part about football for me, in so many ways, is that it's the ultimate living in the moment," Tom said finally. "It's a time capsule. It's a little chess match, and you're volleying the ball back to each other: their offense, our offense. You know it's not going on forever, and I think that in so many ways, you're enjoying that moment, that ebb and flow."

All of this . . . from a curl route. It doesn't take a genius to figure out that Brady was talking about something much greater than just football there. His lifetime of practice all condensed into one throw that was, as he put it, "a masterpiece." Think about how he described that play. He didn't think. He just let the ball fly. He felt like he could do no wrong.

What Brady experienced in that moment was flow state.

//////

I thought about that conversation a lot, wondering if other athletes had moments like Tom's—parts of their career where they constantly wanted to return. I conceptualized it as a show: What if we asked the best athletes in the world to walk us through their greatest moment and tell us what they felt then, what they think about remembering it now? If we gathered all this footage and watched it together, I thought, then somebody could distill all of the responses down into a grand unified theory of athletic excellence. Maybe from these athletes' lives, we could all learn a little something about how to live our own. We dubbed the series *Greatness Code*.

We put together a list of the best athletes in every sport and started to organize interviews. A funny thing happened, the more of these conversations we had. We spoke to Olympians, MVPs, and 100 percent certified GOATs, and they talked to us about world championships, world records, and smaller moments during practice. They came from different backgrounds, played different sports, and were different ages. Some had an answer in mind right away, while some had to think about that question—"What is your greatest moment?"—for a little bit.

But when they started talking, every one of them started to describe a similar experience. They spoke about time slowing down. They spoke about their thoughts becoming clear. They spoke about feeling connected to their sport.

It turns out, the secret to greatness isn't that secret at all.

The secret to greatness is flow state.

Alex Morgan, the superstar striker for the U.S. Women's National Team, said that her greatest moment was in the first game she ever played for Team USA. Morgan came into the game highly touted as the next great U.S. soccer star. The Yanks were playing Canada, in the championship match for the 2012 Olympic qualifiers. Four minutes into the game, Carli Lloyd passed the ball down the field to forward Abby Wambach. Wambach headed the ball forward to a streaking Morgan, who touched it once, twice, and then flicked it into the back of the net. Her first start. Four minutes in. Already, Alex Morgan was scoring goals.

What's incredible watching that play now (and what has somehow become almost commonplace watching the USWNT over the past several years) is Morgan's anticipation. While watching a replay of that goal, color commentator Julie Foudy exclaimed, "Alex started running before Abby even touched the ball!" And indeed, she's right: before Abby Wambach's through ball is on its way, Morgan is already running toward it. "I feel like I'm on the same page with Megan Rapinoe and Abby Wambach and always know what they're going to do, what they need from me, what I need from them," Morgan said. "When you're in rhythm with each other, it's like you have the same pulse."

"When you get into this state," Morgan continued, "everything is so easy." She paused for a moment, then chuckled. "Like, I don't know how to explain it other than it just feels easy."

I spoke with Kelly Slater, the eleven-time world champion surfer. He said that his greatest moment was at Pipeline on Oahu's North Shore, the final tournament of the 1998 season. Slater was surfing to win his fifth championship in a row, a feat that he described to me as being his "lifelong dream." The world standings had been especially tight that year, but Slater's two biggest rivals fell off their boards during their heats. If Slater could ride one good wave, he'd win. And what happened before he paddled out into the water? "I had a full-on breakdown," Slater said.

He started hyperventilating. When a friend walked up and asked, "How are you doing?" Slater just started to cry. The pressure, the emotion, it all felt like too much. I asked Slater what the pressure felt like at that time, and the answer he gave was one of the most insightful descriptions of competition I've ever heard. "Sports define you for a period of time," he told me. "You know what you have to do. There's usually another person or a score you have to get, and so you *have* to accomplish that feat during that time. So, you *know* if you're in the flow, if all your preparation has built up the confidence and the skills you need."

Sports define you for a period of time. What a fascinating way to view athletics. There's great pressure, but also so much opportunity wrapped up in that statement, too. *Are you going to be great?* It's no wonder why the pressure of those types of expectations felt crushing to Slater at the time. But what is it that Slater says he looks for, what is that he says you need to feel in order to know that you're ready to accomplish something great? *You know if you're in the flow.* When Kelly Slater is at his best, he's seeking out flow state.

What happened that day? Slater calmed down, and when he got on his board and started paddling, he says (and pardon if this is starting to sound familiar), "Everything seemed like slow motion." He waited for his wave. He stuck his head in the ocean, dousing himself in holy water so that he could soak in the absolute silence. Then Slater saw a wave, started to paddle in, and won a championship.

"When the greatest moments of your life happen," Slater said, remembering that day, "they don't necessarily happen by chance. It's really an accumulation of all your understanding and knowledge of that thing coming together in that moment. In those times, you don't have to think about it. You become it."

He said that that particular ride felt like the culmination of something to him. "You think about when you were a little child and the feeling you had when you first did this thing: the first time I did a turn or went through a tube or surfed with my dad or my brother. I just remember this flood of different memories coming through, this whole lifetime building to this crescendo."

How could this moment that Slater is describing be categorized as anything other than a religious experience? He's connecting with some larger power, something that unites his past with his present, his soul with a higher being. Surfers aren't the only ones to find something holy in the ocean. The second verse in the Bible, Genesis 1:2, reads, "the Spirit of God was hovering over the waters." Hundreds of years ago, when swells were too big for humans to ride, Hawaiians believed that the gods were riding those waves. Scientists have performed brain scans on surfers while they are riding a wave and have found that almost all of their mind is activated—even more than a comparable scan of a mathematician working out a difficult problem. Other researchers have identified water as being a uniquely suited environment for lulling people into flow state, as it causes people to leave all distractions behind and be rocked gently into focus.

I posed Slater another question: "Why does surfing matter?" He didn't hesitate. Clearly, this is something he had thought about before. "We're riding something that the forces of nature and the universe are just throwing at us. The feeling of riding away, tapping that energy that's there—" He paused for a moment. "I'm not a religious person, [but] that's my church. My church is surfing. There's nowhere else in the world that I get that feeling from."

One more example. We met LeBron James in a hotel room in Sacramento for his interview. The thing about LeBron that most people don't realize is how thoughtful he is. If you were to ask me who the two clearest thinkers I've ever met are, I'd first say my father and then maybe LeBron James.

On this day, LeBron had a smile on his face from the second he walked in. He was laughing to himself about something, but I didn't know what. Then we got the cameras and lighting ready, sat down, and I asked him my question. "What is your greatest moment?"

LeBron started to tell the story of Game Six of the 2012 Eastern Conference Finals against my beloved Celtics. LeBron knew about my Celtics fandom and seemed to be enjoying his ability to rub in the loss one more time. I have only recently forgiven him (not really).

Some context for that game: Heading into that contest, James was the most hated man in the NBA and had (unfairly) gained a reputation as a choke artist. He had put a spotlight squarely on himself two summers before, by joining the Miami Heat via the made-for-TV spectacle known as "The Decision" and then losing in the Finals the next season. Now his Miami team was down 3–2 to the Celtics and they had to travel back to Boston and win on the Celtics' home court if they wanted to keep the season alive. A loss and the famed "Big Three"—James, Dwyane Wade, and Chris Bosh—might be breaking up with not one, not two, not three championships. With zero.

At the time, I remember feeling pretty good about the Celtics' prospects, but you know what happened next. LeBron scored 45 points, with 15 rebounds and 5 assists. There's a defining image of James from that game: hands on his knees, bent down, with eyes looking straight ahead. "He was primal," Miami Heat coach Pat Riley told *Sports Illustrated* of that pose. "He was a cobra, a leopard, a tiger hunched over his kill."

What was he thinking? What was going through his mind that day? LeBron remembered getting on the plane to go to Boston before the game, and he stroked his beard and furrowed his eyebrows. "I felt nothing," he said. "I felt absolutely nothing."

He continued, "I played in Boston so many times during the playoffs, during the regular season, I always felt like this is tough . . . but for this one time, I went to Boston for this Game Six, I felt absolutely nothing."

LeBron said that he barely said a word during the day before and the day of the game. He listened to music, the Wu-Tang Clan, DMX, Jay-Z's "Reservoir Dogs."

"That," LeBron said of his performance that night, "was the result of feeling nothing. I wish I could bottle nothing up, I'll tell you that."

As much as I revile that memory of LeBron torching my favorite team, I equally love LeBron's depiction of it. The underlying DNA of his greatest performance? Nothing. Just a blank void from which greatness sprouts. It's downright spiritual.

Think about all three of those examples. You have three of the best athletes in the world describing their greatest moments, and what do they remember? Alex Morgan felt like she and her teammates had the same pulse. Kelly Slater felt the confluence of every other time he had ever surfed. LeBron felt totally calm on the biggest stage of his career.

They're all experiencing flow. They all spend their careers chasing the ability to get in the zone. All the training, all the practice, all the sacrifices are nothing but offerings to try to get a little closer to that place, to have another moment feeling that way. And when they found it, when they finally did have that moment of transcendence, they accomplished things that few humans can ever dream of doing.

All religions give their followers different tools to tap into those feelings—but only sports gives us all the ability to bear witness.

///////

When we started filming the second season of *Greatness Code*, the first athlete we visited was Lindsey Vonn. Vonn is one of the greatest skiers of all time. She was the first American to win a gold medal in the downhill (when she crossed the finish line, microphones captured her joyous scream of "YES!" and broadcast it across the world in one of those magic Olympic moments), and Vonn was also one of only three women ever to win four World Cup championships—including three in a row from 2008 to 2010. She was so dominant, and her talent so complete, that she won World Cup events in all five skiing events: downhill, super G, giant slalom, slalom, and the super combined. In other words, if you put Vonn on a mountain, she could get down it with more skill and speed than just about anybody else alive. She had her fair share of moments that the whole world could agree would qualify as great.

The latter half of her career, though, seemed to be one setback followed by another. At the 2013 World Championships in Austria, she landed a jump a little off balance, somersaulting down the mountain in a scary crash. A helicopter airlifted her to the hospital, where she underwent surgery to repair her ACL and MCL, as well as a fracture of her tibia. She was never the same skier after that, fighting back to the slopes again and again, only to be dealt another cruel hand of fate. The season after coming back from the knee injury, she tore the ligaments again. She fractured the opposite knee. There was a hairline ankle fracture. Vonn missed the 2014 Olympics with injuries, but she fought back, wanting to make it to one more game before retirement. But in the lead-up to the 2018 Pyeongchang Olympics, she crashed and fractured a bone in her arm, and for weeks she couldn't even lift her arm to brush her teeth. The fracture left her with nerve damage, and when her arm got cold, she couldn't hold her pole. Not to be deterred, Vonn had her coaches duct-tape her pole into her hand. She was going to race anyway.

Why was she so hell-bent on making the Olympics? One, she still wanted to prove to everyone that her 2010 gold medal was no fluke (even though nobody dared to question her greatness). Second, the setting of those Games, in the hills of Korea, held a special place in Vonn's heart.

Years before she was born, her grandfather helped build a ski hill in Wisconsin. It's where Vonn eventually took her first turns. After days on the mountain, her grandfather would tell her stories of his time as a member of the Army Corps of Engineers, when he served overseas during the Korean War. As the Pyeongchang Olympics approached, Vonn was excited to compete in the country that was so important to her grandfather. He made plans to watch stateside.

But just three months before the opening ceremony, Vonn's grandfather passed away. She was devastated. But when she traveled to Korea, she carried a bit of her grandfather with her. "I know he's watching," she said during a press conference when she got to Korea. "And I'm gonna win for him." Around her neck, Vonn wore a locket filled with her grandfather's ashes. A few days before race day, on a rock near the finish line, she scattered them. He would be there with her after all.

In spite of all the injuries, Vonn was the favorite in the race. After all, she was still *Lindsey Vonn*. "I've never felt so much pressure in my life," she told us. When Vonn stepped up to the starting gates, Sofia Goggia of Italy held the lead. "I always prefer chasing somebody. If you know somebody is leading by a good margin and you have to beat that person in order to win or to win a title"—she shrugged before finishing her thought—"it's easy. It's all or nothing."

When Vonn skis, she barrels down the mountain making hairpin turns as fast as 80 miles per hour. Vonn doesn't feel like she's traveling faster than a car on a highway, though. "Similar to *The Matrix* when all the bullets are flying by and everything seems slow, that's how I feel when I'm racing," she said. "I'm in this vortex of speed and time and everything I had learned, everything I had used, everything I was going through, every injury, every obstacle—I use that to get down the mountain."

I asked her what she remembers, specifically, from that run in Pyeongchang. "If I can remember what I'm thinking during a race, I probably didn't win," she said, laughing.

I tried a different approach. "What does skiing represent for you?"

"For me, skiing is a release of all that energy. It's a way of expression and meditation at the same time."

Some people need perfectly dark rooms to meditate. Lindsey Vonn needs a double black diamond. Hey, whatever works, right?

It worked for Vonn that day. For the minute she was on the mountain, she forgot about the pain in her arms and knees, about the pain in her heart. She just skied, like she had years before with her grandfather on his hill in Wisconsin. When Vonn got to the bottom, she missed first place by less than half a second.

Before the day was over, a Norwegian knocked Vonn to third. The greatest skier of all time was going home with a bronze medal. For many, it seemed like nothing short of a disappointment.

But when I asked Vonn about her greatest moment as an athlete, that's what she pointed to. Lindsey Vonn has won gold but she believes her greatest is the one in which she finished third. "Greatness isn't always—and is usually not—picture-perfect," she said. For Vonn, that race was the culmination of a lifetime of working on her craft and refusing to let anything—even her own body—get in the way of her dreams. Third place felt like first that day, because she could feel the convergence of so many aspects of her life with every turn. As she sped down the mountain, she felt freer and freer with each moment. The pain went away. She felt at peace. She felt like a champion.

"And I think my grandfather would be proud," she said.

//////

As I was reviewing the interviews with all these athletes, I started to consider my own experience with flow. I've had moments while editing documentaries and while writing when I've felt everything else slip away and slow down. I've felt it while jogging and while shooting hoops. But along with all these experiences, I also found my mind drifting toward the night I sat in the Boston Garden as a kid, witnessing Michael Jordan dazzle against my Celtics.

Remember, back in chapter one, when I described that moment? MJ could not be stopped that night, scoring every way that somebody could score on a basketball court. Larry Bird described Jordan as playing like God, and his comment made me start thinking about the ideas that one day turned into this book. I found it odd that this moment, when I sat passively in the stands watching a rival, would come to my mind when thinking about flow state. But that's exactly what happened. Certainly, MJ was in the zone that night, but the more I thought about it, the more I realized that I experienced it, too: me and MJ and all the others watching the game live and even those who pull the game up today and watch him on YouTube—we all felt it.

Sitting in the Garden that night, I stopped glancing at the clock. I stopped going through the wild highs and lows that come with play-off basketball. I just watched, as if I was going through the motions of being a die-hard Celtics fan. All around me, it seemed like the magic that filled the building that night had enchanted everybody there. We all watched together. We all felt MJ's energy. We were one. Every one of us was in the zone.

That, to me, is something rather powerful about the nature of sports. Not only can we go play catch, shoot hoops, or go for a run, and feel something similar to the inner peace that somebody can find during intense meditation or prayer, but through sports, we can actually take on the experience of others. One person's flow state can be transferred to a teammate, and then to a stadium, and then to an entire country watching together. We can all share its benefits—the happiness, the calmness, the activation of the smile muscles. We can all be connected to something greater than ourselves.

Bodhidharma's puzzle, as he journeyed through China thousands of years ago, was that he needed to get in front of a group and make everybody feel exactly the thing that he was experiencing. But he struggled to discover a method to do so with the right type of vitality. When he simply tried to meditate, he couldn't find an eager audience. It's no surprise, then, that Bodhidharma eventually turned to sports. What he needed to do, all along, was get in the zone.

LINDSEY VONN

NICKNAME: The Fastest Woman on the Planet

DESCRIPTION: American Skiing Legend

FACT #1:

In 2010, Vonn became the first American woman to win a gold medal in Olympic downhill skiing.

FACT #2:

In races, Vonn sometimes reached top speeds of nearly 90 miles per hour.

WISDOM:

"When you fall, get right back up. Just keep going, keep pushing it." —*Lindsey Vonn*

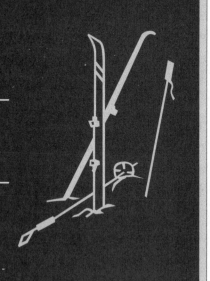

CHAPTER 5

Moral Codes

Sure, people feel community around sports. But does sports, like any religion, teach its followers about the difference between right and wrong? Can sports teach us how to live our lives?

Learn to play the game of life, as well as the game of athletics, according to the rules of society. If you can take that and put it into practice in the community in which you live, then, to me you have won the greatest championship.

—Jesse Owens

THE POPE WAS worried. It was the 1880s, and as an unprecedented wave of Catholic immigrants moved to the United States, American culture started to take hold over the first generation of American Catholic children. They listened to American music, preferred American foods, and even started playing baseball. It seemed, for a moment, like the church's powerful hold on their everyday life was loosening.

All over America, in churches and around the dinner table, Catholics spoke in hushed tones as they discussed the issue. What could be done to preserve their beliefs? And who was to blame? Soon a consensus emerged. The culprit had to be the American public school

system. Catholics believed that these schools, Protestant in tempera-
ment, threatened to lead their children down the wrong path. Those
schools were why their kids couldn't follow along during Mass, had
never learned the prayers, and didn't behave. It's why they couldn't
even recite Bible verses by heart.

From those small meetings sprang bigger meetings and bigger
gatherings, and the debate about what should be done grew more and
more impassioned. The discussions soon rose up the Catholic food
chain until the news traveled across the Atlantic and all the way to the
Vatican. That's when the pope consulted with his advisers and prayed
for an answer. Soon he had one. "There can be no ideally perfect education which is not Christian education," Pius IX declared. "Every effort
must be made to increase the number and quality of parochial schools."

That news traveled back over the ocean and spread from church
to church. Soon all the bishops in America met to formulate a joint
response. Out of that meeting came a decree. The bishops called for
every parish across the country to establish its own school. This task,
they said, would be an "absolute necessity." Further, in order to instill
the proper virtues in the next generation, every Catholic parent would
be required to send their children to a Catholic school.

The schools popped up one by one, all across the country. By the
turn of the century, 3,500 Catholic elementary schools had been established across the country. By 1920, that number doubled. Catholic
nuns taught nearly two million students a year. By the 1960s, Catholics educated 5.5 million students annually with affordable, quality
schooling.

Faith touches everything in a Catholic school. That's by design.
It's kind of the point. As Edward A. Pace, the former vice rector of the
Catholic University of America, once explained, Catholicism must be
"the foundation and crown of the youth's entire training at every level
of instruction."

In other words, everything from math to geography should teach students about the gospel.

I never went to Catholic school, but I did grow up in Boston—which is to say that I saw the impact that the Catholic Church had on nearly every aspect of the city. Being a kid in Boston also meant that I had tons of friends who came up through the Catholic school system, and as I'd meet them on the weekends for Little League games, we'd talk about the characters and teachers that made our schools unique. Soon enough, I started to learn about the stubborn nuns and strict guidelines.

My friends also told me about a dreaded class known as religion. Coming from my particular home, it sounded familiar, a dedicated period of the day set aside for wide-ranging discussions about spirituality and the secrets of the universe. But that, my friends assured me, was not what religion class was like. Not at all. Religion class was dry as can be. The history of the church. Reciting the Apostles' Creed. Memorization followed by more memorization with a little more memorization after that.

One of those friends was stressed one day when we met, worried about an upcoming quiz. He couldn't remember the right order of the Ten Commandments, always flipping "You shall not kill" with "You shall not commit adultery." He said he would fail if he got the order wrong. We were young, and I asked him what *adultery* even meant. He had no idea—but that wasn't really the point.

What *was* the point of it, then? It was a question that turned over in my mind at the time. The purpose of the school, clearly, was to model their faith. But instead, it just seemed like it was training my friends to go through the motions. Most of them didn't think about what prayer they were saying. It was just a series of words, recited in a specific order. The lessons they learned seemed rather empty. The same boys who sang in the choir hosted the biggest parties during high school. Clearly, not everybody was taking classroom knowledge and applying it to everyday life. Some got it. Others didn't.

But everybody understands the lessons of the Religion of Sports.

//////

At the same time that Catholics built their school system in America, sweeping societal changes were occurring across Europe. In France, industrialization transformed towns into modern cities, making some men fabulously wealthy and others unbelievably poor. Émile Durkheim came of age in this new world, born in 1858 near the French-German border. His parents raised him Jewish, but Durkheim didn't practice any faith as he got older.

Instead of practicing any religion, Durkheim studied all of them.

Today Durkheim is considered one of the creators of modern social science. With the world changing around him, he grew obsessed with the ways in which people interacted with each other. His work led to our modern understanding of mass culture, and in one seminal work, Durkheim turned his gaze toward religion. Why, he wanted to know, did humans seem drawn to seek out spirituality?

To do his analysis, Durkheim studied Aboriginal tribes in Australia to see how they interacted with their faith. If he could understand how they viewed the spiritual forces in life, he believed, he could then generalize the findings to all cultures. Despite the fact that such a study could never be commissioned today, it still remains one of the foundational texts in modern religious studies. Durkheim concluded that religion constituted one of the fundamental elements of humanity.

What *was* a religion? Durkheim identified three fundamental elements that every religion must share: a dichotomy between the sacred and the profane, a set of beliefs and practices, and a clearly defined set of morals, morals that help shape what entire societies consider right and wrong. His thoughts still influence religious scholars today.

Durkheim had another passion as well. For decades, he was a teacher. He first taught at French high schools and later at a university in Bordeaux and finally at the famed Sorbonne in Paris. As a professor, he lectured on numerous topics, but he taught education first and foremost. Durkheim studied culture, saw how it was changing, saw how religions played a role in that society, and grew obsessed with understanding how we mold future generations. Education, to Durkheim, was "the methodical socialization of the younger generation." He believed that younger generations could best be understood as "soft wax." It was the education system's job to mold those children—but what would be the best way to do it?

Durkheim taught a class that tried to find an answer to this question. It was called, very originally, "Moral Education," and he would lecture in front of huge groups of France's future teachers. For Durkheim, how to instill a moral code into the youth was a question of national importance, and he consulted with the French government to reform the standardized curriculum in the country. At the heart of his course lay a conflict and contradiction: according to Durkheim's own beliefs, establishing a clear set of morals was a foundational element of a religion. Creating a moral code was one of the things that religions did best. But Durkheim didn't think that religions were particularly good at teaching those morals to children. Religion was too indoctrinating, too rigid for children to understand their lessons. Kids couldn't simply be forced into a mold. They were soft wax, after all, and they needed to be slowly and carefully formed.

Morals, he believed, were best learned through experience, through the existence of a "moral environment." Rather than reciting the Ten Commandments, for example, kids would need to exist in a space that personified those values if they were to internalize them. If hard work is a value that's encouraged, then hard work would need to be demonstrated and rewarded.

How did Durkheim plan on implementing this? His answer was high on inspiration for his audience of future educators and low on specifics. It was, in some ways, a bit of a cop-out. A teacher's job, Durkheim explained, was "to prepare the future." Therefore, they would have to create those moral environments. As for how they could best achieve this, Durkheim left the question unanswered.

Reading Durkheim today, it feels like some of his work on education and morals is unfinished. Religions create morality but aren't equipped to teach it. Teachers must instill those morals but aren't given instructions on how to actually do so. Isn't there a better way?

Enter, once again, the Religion of Sports.

//////

About as soon as my son, Krishu, could walk, I started to sign him up for sports. Tee-ball, basketball, flag football, tennis, soccer, karate ... It wasn't part of some long-term plan to turn my son into the next Tiger Woods, winning golf tournaments before his tenth birthday. Like any parent, I wanted Krishu to learn respect, accountability, and hard work. I turned to sports, more than anything, to provide him with a moral education.

Sports, like any religion, has a moral code. And even though we've focused thus far only on sports on their largest scale—the NBA Finals, the Boston Marathon, the Olympics—some of the most fundamental truths of sports can best be seen on the smallest of stages, on playgrounds and in Little Leagues across the world. Through youth sports, we create a world in which kids can viscerally understand many of the most important lessons we can teach them. We create an ideal version of Durkheim's "moral environment." Win or lose, opponents shake hands. The lesson: humility. Cheating is never tolerated. The lesson: integrity. Referees enforce the same rules for everybody. The lesson: fairness.

There are so many more examples. Accountability. Resilience. Leadership. Discipline. Teamwork. Sports teach us how to survive and adapt, how to push ourselves and discover our highest potential. They also help solve Durkheim's puzzle. They are a source of morals, as well as a venue in which we can actually experience them—and therefore learn.

Krishu was hesitant at first. I remember him holding on to my pant leg and refusing to go on the floor at karate class. He was shy around others, frustrated that he was never the best. When Krishu was seven, we started to hear from friends about a taekwondo teacher, a *sabom*, who had a studio nearby. His name was Master Quan. "He's amazing," friends at school told us. "There's nobody else like him."

My wife, Candice, and I signed Krishu up for a beginner's class. I had never practiced taekwondo as a kid; I honestly knew next to nothing about it. Candice and I lingered off the mat with other parents, watching as Krishu joined the rest of the group. An American flag hung next to a South Korean flag on the wall. Pads and punching bags were stacked in the corner. Master Quan towered in front of the group of kids, wearing a black suit with a black belt. He was younger than I was expecting; not some old Mr. Miyagi type but a tall, strong, and handsome thirty-year-old. He bowed to the group. A few of the kids didn't return the gesture, including Krishu. Master Quan looked toward these dissenters. "Before you can do anything else in taekwondo, you have to learn respect," he explained. "You bow to your teacher, and you bow to your opponent." Then he bowed his head toward Krishu, and Krishu bowed right back.

His moral education was just beginning.

///////

Coaches have long existed as legendary, wise figures in our culture. Turn on a TV and see *Friday Night Lights'* Coach Taylor, preaching the mantra of "Clear Eyes, Full Hearts, Can't Lose." Watch *Hoosiers* and see Coach Norman Dale proselytize on the importance of effort over outcome. There are so many more: Herman Boone from *Remember the Titans* ("I don't care if you like each other or not, but you will respect each other"), *The Mighty Ducks'* Gordon Bombay ("A team isn't a bunch of kids out to win. A team is something you belong to, something you feel, something you have to earn"), and more recently, *Ted Lasso* ("Takin' on a challenge is a lot like riding a horse. If you're comfortable while you're doin' it, you're probably doin' it wrong"). When Hollywood needs someone to communicate the difference between right and wrong, more often than not they turn to a coach to deliver the message.

That reputation is well earned. For so long, coaches have represented the best of us, the person who can see what others overlook, who can push us to become the best possible version of ourselves. They are sports' priests, rabbis, and imams. As the legendary Alabama Crimson Tide football coach Bear Bryant once said, "Mama wanted me to be a preacher. I told her coachin' and preachin' were a lot alike."

Here's one example: In 2019, Liverpool won eighteen matches in a row in the Premier League. Daragh Curley, a nine-year-old Manchester United fan from Ireland, wasn't happy about it. He wanted the Reds to stop winning, and so he wrote Liverpool manager Jürgen Klopp a letter. "Liverpool are winning too many games," he said. "Being a United fan that is very sad. So next time Liverpool play please make them lose. You should just let the other tam [*sic*] score."

A few weeks later, Daragh's parents visited the post office when they were told that a letter from Liverpool had arrived for their son. It was written on official Liverpool FC stationery with Klopp's signature on the bottom, and it read, "As much as you want Liverpool to lose, it is my job to do everything that I can to help Liverpool to win as there are millions of people around the world who want that to happen, so I really do not want to let them down."

Klopp went on to say, "Luckily for you, we have lost games in the past and we will lose games in the future because that is football. The problem is when you are 10 years old you think that things will always be as they are now but if there is one thing I can tell you at 52 years old it is that this most definitely isn't the case."

And finally, "I hope that if we are lucky enough to win more games and maybe even lift some more trophies you will not be too disappointed, because although our clubs are great rivals, we also share a great respect for one another. This, to me, is what football is all about."

This letter, to *me*, captures the meaning of coaching. Even through a letter, Klopp takes a young man and shows him the way to lead a meaningful life. Klopp teaches Daragh empathy, the fleeting nature of success, and sportsmanship. It's extraordinary, really.

Other coaches distill their philosophy into catchy and easy-to-digest metaphors. Take University of Minnesota Golden Gophers football coach P. J. Fleck, for example. In 2012, Western Michigan University hired Fleck and made him the youngest coach in major college football history when he was only thirty-two years old. How'd he get to that point so quickly? His secret lies in his philosophy. For Fleck, it's all about rowing the boat.

Look at those Western Michigan teams or Fleck's Gophers today and you'll see the phrase "row the boat" everywhere. It's on the walls of locker rooms. It's on T-shirts. When Minnesota makes a big play, their mascot, a cartoonish smiling creature known as "Goldy Gopher," holds up an oar with the letters *RTB* painted on the side. When the team breaks the huddle, what do they all put their hands in the middle and say?

Row the boat.

Fleck says the phrase came to him right before that first coaching job. In 2011, Fleck's son passed away suddenly from a heart condition. "As you hold your son as he passes away, your whole life changes," Fleck explains. "What you believe in, how you've believed in it, what you've done to that point, where you're going to go and how you're going to live your life all changes. . . . It's a never-give-up mantra that has to do strictly with life or adversity or handling success, never giving up."

He sounds like he's delivering a Sunday sermon. The way he explains it, you need three things to row a boat: the oar, the boat, and the compass.

The oar represents the energy, passion, and work you bring to your life. It's the only thing that actually moves the boat.

The boat is the sacrifice. If you sacrifice a lot, the boat will be sturdy, and the crew can make it through any obstacles. If you don't, a simple obstacle can capsize everything.

Finally, you need a compass, which represents the direction you're trying to go. To get there, you need to surround yourself with people who are all trying to go in the same direction.

There's one more element to the metaphor, one that many of Fleck's players say is especially key to their success. When you're rowing a boat, you're facing backward, unable to see what lies ahead. You just look behind you, toward your past, and as you study it, you learn from it. Then you just keep rowing and have faith that you'll get to where you need to be.

Fleck learned and developed all those lessons *off* the football field, but it's telling that football is the vessel through which he thought they could best be translated. He chose coaching as the best venue to spread his message. That calculus makes sense. Who else but a coach could preach about rowing a boat and turn it into an entire university's rallying cry?

Of course, there was no greater mastery than the great John Wooden, the bespectacled architect of UCLA's basketball dynasty. Wooden coached the Bruins from the 1940s until the '70s, and is generally regarded as the most successful coach in basketball history.

When he took over at UCLA, the basketball program was middling, playing in a small gym without many resources. That all changed under Wooden. During one twelve-year period with Wooden on the sidelines, UCLA won 10 national championships and 88 consecutive games.

He won not with particularly brutal practices of wind sprints and powerlifting but with doctrine. Winning wasn't the only thing . . . it was the accidental by-product of maximum effort. "Championships were never the cake," Wooden once said. "They were the icing. Doing our best was the cake." On a website created to honor their father's legacy, his family wrote that "[t]he gyms he coached in became chapels, the court a pulpit where he preached a different kind of success approach."

Wooden earned the nickname "The Wizard of Westwood," but what he was doing wasn't any type of magic. In many ways, he's the archetype of everything we have come to value in a coach. (When Jason Sudeikis developed the character of Ted Lasso, he recalled his Wooden-quoting high school basketball coach.) Wooden simply used basketball as a way to instill the values he found to be the most important in living a meaningful life.

On Wooden's wall, from the first day he arrived at UCLA until his retirement, hung a hand-drawn version of what Wooden called "The Pyramid of Success." He drew it as a twenty-four-year-old high school basketball coach in Dayton, Kentucky. Coaching that high school team, Wooden started to develop his philosophy on coaching. He thought that he could lead any team to victory not with superior schemes but by creating a team that was prepared for life as much as they were for their next game. He formulated his theories into a pyramid.

Pyramids, I should remind you, were originally constructed in Egypt as holy, religious sites. But I digress.

His theory was that at the foundation of any path to success were five traits: industriousness, enthusiasm, friendship, loyalty, and cooperation. From there, more traits could be stacked on top: self-control, initiative, team spirit, poise, confidence. The whole thing was held together with Wooden's "mortar," qualities like ambition, sincerity, honesty, and patience. At the top, only able to be reached if all the other qualities are practiced, was competitive greatness.

Even when it did come to specific basketball strategy, Wooden found ways to incorporate his pyramid. His teams valued quick ball movement, pass after pass after pass. "Move!" he'd holler from the sideline. "Move! Move!" The lesson placed an emphasis on teamwork and trusting your guts. When asked to describe his team's style of play, Wooden once said, "Be quick, but don't hurry."

I've always thought that that is a pretty great way to live a life.

Years after he retired, Wooden adapted his pyramid into a series of devotionals for Bible studies. Wooden, after all, went to church nearly every Sunday. But even with that background, he didn't immediately turn toward the church to spread his beliefs about morality. Basketball was his first choice to spread his gospel. After all, sports are the place where we, as a society, have decided to forge character.

///////

Krishu came to look forward to his time at Master Quan's studio. I'd go to pick him up from school, and he'd run out practicing his punches and kicks and tell me how excited he was for the next class. Quan gave Krishu goals. If he stuck with it, Quan said, Krishu could become a black belt. But if he really wanted to do so, it would require discipline, hard work, and lots and lots of practice.

First, he suggested to Krishu, start with earning a yellow belt. To get that, he'd have to begin to master some of the technical aspects of the sport: the proper stance, legs spread a little more than shoulder length. Fists had to be held tightly before practicing a punch. Knees should point at the target before firing a kick. Just as important was the etiquette. After the initial lesson on bowing, any student who entered the studio, known as a dojang, was to acknowledge Master Quan, along with the other Masters, or else they'd be forced to leave and try again. Their uniforms were to be clean , and belts needed to be tied correctly.

One day, Krishu was upset when I picked him up from the dojang. Other kids had already started to break boards, he said. He wanted to break boards, too. I asked him why he wasn't allowed to try it yet. According to Krishu, Master Quan told him that he simply wasn't ready. The others had practiced harder, Master Quan said. He could tell.

"Well," I asked Krishu, who was several seasons into a Madden franchise at the time, "is he right?"

Krishu didn't say much else for the rest of the drive home, but when we got there, he went straight upstairs into his room. I was expecting to hear the sounds of crying, but instead I heard the ground shaking as he practiced jump kicks and the soft thud of punches hitting the pillows on his bed. He was taking Quan's lessons to heart. Two weeks later, Krishu asked me to stay to watch him practice.

In the middle of the session, Master Quan brought out a wooden plank. Krishu stood in front, feet a little wider than shoulder width. He made his hand firm and stared down the wood as if it would strike him back. Then he hit it just like he practiced, hard and fast, and the wood shattered into two. He bowed to Master Quan, and then looked at Candice and me and smiled.

I never expected Krishu to stick with taekwondo for more than a few lessons. But soon it had been a year and then another and then another. He loved it. He loved the structure it gave him, the way that it provided him something to work toward. He could see that he was getting better as he earned each new belt. All along, Master Quan pushed him to be better. Not just a better athlete, although the drills made sure that Krishu was increasing his skills in that regard. Quan was teaching Krishu about life through martial arts. He was making my kid tougher, wiser, kinder.

One day, Master Quan came up to me and Candice after class. "I think Krishu's ready for tournaments," he said. "I think he's ready for that new type of challenge."

///////

Many weekends, Candice and I will drive Krishu to some forgotten corner of the Los Angeles area to a high school gym or community center that is the home of that week's taekwondo tournament. Families drive hours so that their children can make the pilgrimage to compete. There are kids of all ages, wearing their white uniforms, called doboks. They have black belts and purple belts and brown belts. They're tall and short and skinny and fat.

The floor of the gym is covered in mats. Several matches go on at once, so you hear smacks and screams and whistles and the beeping of electronic scoreboards. It smells. In taekwondo, each competitor wears headgear and various pads protecting their bodies. You get points by landing kicks and punches—but only if it hits the part of your competitor that is protected. Punches to the torso are one point, straight kicks to the torso are two. Add in some acrobatics and you'll get three points for a spinning kick. Land a spinning kick to the head and you've got yourself four whole points, the maximum for any move. Whoever has the most points after three two-minute rounds wins. Oh, and if you strike your opponent weakly, the points don't count. You have to use force.

As a parent, the whole thing is just a little bit stressful.

These types of youth sport spectacles often breed a particularly nasty type of person: the overbearing parent. There are people who, upon seeing their kid's opponent, will scream taunts like, "See? He's small and weak. You can knock him out!" If a referee awards that opponent an even remotely controversial point, the parent will berate them with insults. "You must be blind!" If their kid loses, the kid is reprimanded in front of everyone, told that they didn't try hard enough, that they sure as hell better not lose the next one, or else . . .

It's important to acknowledge these darker sides of the Religion of Sports. These types of people interpret their faith in sports differently than I do. They're not appreciating the lessons that sports can teach, nor do they see the damage that they're causing. They just push, push, push. They view sports as a means to an end, as a path to a scholarship or Olympic berth. Imagine what they could learn if they just let the sport work its magic. This type of mindset is its own special type of sporting sin.

I get it, though. It's tough to stay on the sidelines.

One weekend, we drove out to Oxnard, where Krishu fought a couple of matches and looked strong. He won some and lost some others. Then, up walked his next opponent. This kid was huge. As he warmed up, he executed moves that Krishu had never seen before. It was like when Daniel from *The Karate Kid* faced the kids from the Cobra Kai dojo. I saw Krishu's eyes go huge, and all I wanted to do was run out there and grab him. Candice and I knew he was going to lose. Krishu knew he was going to lose. What was the point of going on?

Master Quan approached Candice and me. He seemed to know what we were thinking. "Don't say anything to him," he told us. "Just let him finish the match." Then Master Quan pulled Krishu aside and spoke to him for a minute. Krishu didn't say anything. He just nodded and nodded and slowly turned around, walking right up to the big kid. He bowed, then prepared for the match.

I wish I could tell you about some miraculous performance. I wish it was actually like *The Karate Kid*, Krishu channeling some never-before-seen potential where he shocked everyone in the gym and won the match. But that's not what happened. Not even close.

I bit my tongue and watched the bigger kid land punches and kicks and just absolutely dominate. Krishu never had a chance. He was hurt—not physically, but certainly mentally—and after he bowed, he hurried off the mat to Master Quan, who led him outside.

I gave them a minute before following. They were alone on the far side of the gym, and Krishu was crying hard. Candice and I both gave him a hug and told him he did his best. I told him how proud we were of him. Master Quan looked at me and gave me a thumbs-up, and so I gave them their space again. From afar, I could see them having a long talk. Krishu's crying slowed down. He got that quiet look of determination again. He gave Quan a fist bump and went back inside to cheer on one of his friends.

On the way home, I asked Krishu what Master Quan told him before the fight. "He said that I was probably going to lose," Krishu told me. "But he said that was okay. He said that every time I got knocked down, I had to stand up, and that no matter how much I wanted to cry, I had to hold it all in until it was over." He paused for a second. "And then afterwards he was telling me how proud he was of me. He said that if I never gave up like that, then one day, I'd be able to beat a kid like that."

I've been on the field for Super Bowls, in locker rooms after the NBA Finals, and in crowds with hundreds of thousands of roaring fans. But without a doubt, that's my favorite sports moment I've ever been lucky enough to witness.

Lessons like that capture the unique power that sports have to teach morals. Here is a medium with clear rules, where bowing to your master is as important as executing an acrobatic kick. It's a place where we set goals and see the difference practice can make. It is, in the words of Durkheim, a "moral environment." We don't just recite the lessons. We live them. We experience them. We practice them. Then, we take those lessons and apply them elsewhere. Sports, in no uncertain terms, help teach us the lessons we need to live a meaningful life.

A few days after the tournament, Krishu was working on a math problem for homework. He couldn't figure out the answer during class, so he brought it home to try to figure it out. Again and again, he got it wrong. But it never fazed him. "This problem is like that kid in the tournament," he said.

In that moment, I was more proud of my son than I would have been if he brought home the championship.

JOHN WOODEN

NICKNAME: The Wizard of Westwood

DESCRIPTION: Basketball coach, UCLA Bruins, 1948–75

FACT #1:	FACT #2:
One of the most successful and beloved coaches in any sport; Wooden's Bruins won ten national championships, including seven in a row.	Coached legends like Lew Alcindor (who later changed his name to Kareem Abdul-Jabbar) and Bill Walton, using inspirational messages and his famous "Pyramid of Success" to inspire greatness beyond the basketball court.

WISDOM:

"Be more concerned with your character than your reputation, because your character is what you really are, while your reputation is merely what others think you are." *—John Wooden*

CHAPTER 6

Pilgrimage

If you build it, they will come . . .

I touched the wall, and all my dreams, hopes, and ambitions basically coalesced into one moment.

**—Australian swimmer Duncan Armstrong
on winning a gold medal at the 1988 Olympics in Seoul**

YEARS AND YEARS ago, according to Hindu mythology, the gods and the demons joined forces. The ocean at this time was a frothy, milky mess. Somewhere in its depths, they knew, lay the nectar of immortality—whoever drank even a few drops would live forever. Neither side was strong enough to recover the prize on their own, so they agreed that they would work together on the condition that the nectar would be split between the two groups if it was ever found.

They used a mountain to churn the water, and a snake as the churning rope. The demons held the tail, and the gods held the head. Once, the snake coughed up a drop of venom, and to keep it from contaminating the ocean and, potentially, the nectar that promised eternal life, one of the gods caught the venom in his throat and held it there until his throat turned blue.

As the ocean churned, treasures started to surface. The moon. The goddess of wine. A tree that could fulfill any wish. Finally, there it was: a pitcher filled with the nectar of immortality.

But about as soon as they uncovered the elixir, their agreement to share it equally among themselves dissolved. The gods and the demons fought, and in the middle of the battle, one of the gods, named Vishnu, transformed himself into a beautiful woman. The demons got distracted, and while the woman wooed them, the rest of the gods quickly drank the nectar. That is how the gods have managed to live forever.

But in their haste, the gods spilled. Drops of the nectar of immortality fell onto four spots alongside rivers, in the cities today known as Ujjain, Prayagraj, Haridwar, and Nashik, in modern-day India. The story goes that if the timing is just right, you can bathe in those rivers and get some of the power of that holy water, sins washing away. The idea is that it's never too late to reach a higher being. You can always cleanse your soul, and this particular river in this particular place can help you do so.

A festival has sprung up around the event, which rotates every three years between the four holy locations. It is known as Kumbh Mela, which translates to "pitcher festival." It's the largest pilgrimage in the world—the largest *gathering of humans* in the world, actually, and it's hard to fully grasp the scale of the event.

Hindus travel from all over to cleanse themselves and be closer to their faith. In all, Kumbh Mela lasts around fifty days. In 2013, 120 million pilgrims traveled to Prayagraj to show their faith—roughly equivalent to the entire population of Japan. On the day considered the most opportune for cleansing, a whopping 30 million people attended. There's no invitation. There's no central organizing body. Everybody just shows up.

When the festival is occurring, Hindu priests, known as Nagar

Babas, leave behind their homes (and clothes) for the entire three months of the festival. They cover their bodies in ash to symbolize that we will all die one day, and they're the first to dip themselves into the river. They swim and splash each other in the river, holy men playing like kids on a hot summer day. Then the rest of the crowd follows. For months, at all hours of the day, pilgrims slowly approach the water and dip themselves. Songs play in the background. Monks chant prayers. Everyone was called here at the same time for their own reasons and one universal motive, too: coming to Kumbh Mela is an expression of one's inner journey toward faith.

///////

Every religion has pilgrims. One of the tenets of Muslim faith is the requirement to make a hajj—the Arabic word for, literally, pilgrimage—to the city of Mecca. Up to three million travel to Mecca annually, gathering around the Kaaba, a black cubic structure draped in cotton and silk at the center of the town's great mosque. According to Muslim faith, the prophet Abraham built the original structure with his son Ismail. After it had been built, God commanded Abraham to instruct all believers to make a pilgrimage to the site, which would represent their devotion to God.

Catholics visit the Vatican; 7 million gathered there in 2015 for the benediction of Pope Francis. Some pilgrims walk the Camino de Santiago from France to the northwestern edge of Spain, all the way to the tomb of St. James. The Torah instructs Jews to visit Jerusalem three times a year, and even if that isn't always possible, many Jewish people try to make it to the Western Wall at least once in their lives. The fig tree in Bodh Gaya in India is said to be a descendant of the tree that the Buddha meditated under for seven years, and Buddhists travel there to feel linked to him. Some anthropologists believe that Stonehenge was a site that pilgrims journeyed to in order to heal from chronic illnesses. During times of drought, the Mayans visited the temple at Chichen Itza to offer sacrifices to their gods in exchange for rain.

This idea of traveling to a holy place is one of the most fundamental aspects of any religion. There's something irresistible about its symbolism as well: a believer takes a physical journey and expresses their own faith in the process. The journey toward a holy site then mirrors a journey toward a religious ideal. The same, without a doubt, can be said of the Religion of Sports—and not only because the Milwaukee Bucks used to play in an arena known as "Mecca."

One recent survey found that 43 percent of Americans included going to the Super Bowl as an item on their bucket list—more than the amount who wished to visit Europe. That statistic reveals two things about the nature of fandom: one, we have a deep desire to see our teams reach the Promised Land (the survey doesn't say so, but I bet most people would specify that they hope to visit a Super Bowl that includes their beloved team), and two, physically going to games can be a spiritual journey. If pilgrimages are a manifestation of someone's faith, then aren't the acts of going to a game—from scalping tickets to donning a jersey to crafting the perfect sign—all signs of someone's devotion? We sing a hymn to honor these sports pilgrimages: "*Take me out to the ball game, take me out to the crowd...*"

In sports, as in traditional faith, all pilgrims in a stadium have their own reasons for gathering there. For some, the call to a specific stadium can be a familial bond. A friend's grandparents honeymooned in the bleachers of Wrigley Field; that's why he always tries to sit near left field. Another baseball fan might travel to Cooperstown, New York, supposedly the birthplace of the sport and the home of its Hall of Fame. The entire town is now a shrine to America's Pastime, and the museum is chock-full of relics like Jackie Robinson's cap and Babe Ruth's bat. Sometimes, the pilgrimages are annual traditions, like the hajj: the golf world journeys to Augusta National every April, and Formula 1 invades Monaco each May. Other pilgrimages can be made anytime, like a Catholic's trip to St. Peter's Basilica: 32 million pilgrims toured FC Barcelona's stadium Camp Nou in 2017. It's the most popular attraction in the city, and those tours aren't even occurring on game days. People just want to see the field where so much magic has happened.

That's the fans' perspective—but athletes experience pilgrimages of their own as well. St. Augustine of Hippo, who did as much to influence Christian faith as anyone since biblical times, once wrote that he considered life itself to be one big pilgrimage, a constant journey on the way to the Promised Land. Athletes tend to view their lives similarly—as a series of smaller pilgrimages made while on the road to a larger one. All season is a long path to a championship, a career a climb toward greatness.

That path isn't always a straight one. Just ask the most famous athlete you've probably never heard of: Sachin Tendulkar.

///////

Every summer when I was a kid, my family traveled to India to see my grandparents. It always felt like I was entering a different world—especially because nobody followed the same sports that I did. Most days, I would sneak off and pay to make a long-distance phone call to a service called "Sports Phone." I dialed in and listened to a minute-long recorded message reporting the day's scores. That's how I would track the Red Sox throughout the dog days.

Those trips to India were family time, and my sister and I would spend all day with our cousins. I'd come back from the phone excited about a Red Sox win and tell my cousins why each game meant so much. They always looked at me puzzled.

"Baseball," they'd tell me, "is just a worse version of cricket."

We must have had the same debate hundreds of times. I'd tell them about pitchers, and they'd tell me about bowlers. I'd tell them about home runs; they'd tell me about sixers. I'd ask why a cricket game lasted for days at a time; they'd ask why a baseball game "only" lasted nine innings.

Sometimes they'd take me to play cricket with their friends. I didn't completely grasp the rules, and when they handed me the ball, I hurled it overhand like a fastball. They laughed and laughed, calling me a "chucker." I learned quickly that the term wasn't a compliment.

Cricket is everywhere in the country. "Holding a bat is probably the first thing that anybody does in India," Virat Kohli, the cricket star, told me. Boys play anywhere they can spread out for a game—from busy intersections to fields on the side of highways, like stickball in Depression-era New York. "Painting three wickets on a wall, picking up a rubber ball, and makeshift bats," remembered the Bollywood legend Amitabh Bachchan. "We'd play cricket in the alleys, on the streets."

I never totally embraced cricket, just like my cousins never fell in love with baseball. But those summers ensured that I always kept tabs on the sport even when I went back home to Boston. In 1983, the Indian National Team won their first cricket World Cup. It was the first major international trophy the country had ever won, and the next time we visited, my cousins reenacted every play for me.

But India wouldn't taste that success again for decades. Every tournament in the years following that 1983 victory followed a script that felt familiar to me as a long-suffering Sox fan. They'd get painstakingly close, only to let success fall through their fingertips at the last possible second. Said cricket commentator Harsha Bhogle, "In India, we look at defeat as this colossal monster that's going to strangle you. And it is the fear of defeat that I think produces more defeat."

Yet the Indian National Team had something that no other country had: Sachin Tendulkar, the best player in the long history of the sport. He made his professional debut when he was only fifteen years old and scored 100 runs—called a "century"—in his first game. Sachin became a star unlike any ever seen in the country—perhaps unlike any other in the world. Everywhere he went, the Indian government provided secret service agents to keep the throngs of fans away. There's a well-known book written by Indian authors Vijay Santhanam and Shyam Balasubramanian called *If Cricket Is a Religion, Sachin Is God*. It's been reprinted five times.

In 2011, Sachin approached the end of his career. It was going to be his last World Cup, his last opportunity to deliver a trophy for his country, the only accomplishment he had yet to achieve in the sport. Best of all, the World Cup would be hosted in India that year. Sachin could leave the game as a hero on home soil.

Directly outside the hotel room where Sachin stayed during the tournament was a billboard. It had a picture of Sachin on it and said, "Sachin feels the burden of 1.3 billion people."

No pressure.

///////

Right around that time, ESPN began producing *30 for 30* documentaries. ESPN's Connor Schell and Bill Simmons wanted to spotlight sports from all across the globe and introduce their viewers to characters they might never have heard of before. I thought Sachin and his quest at the 2011 World Cup might be the perfect fit and sent them a pitch.

"Can you get an interview with Sachin?" they asked me.

"Oh yeah," I said. "Totally. For sure. No problem."

I had no connections to Sachin, but that was no matter. How hard could it be? I'd started to get to know Kobe around that time, and that experience had emboldened me. I sold my knowledge of cricket and access to Sachin hard. ESPN green-lit the project, and soon I was planning to take a pilgrimage back to my ancestral homeland to decode a sport that in truth I never quite fully grasped. I quickly realized I needed someone who could help me bridge the gap between my Western sensibilities and India's fascination with cricket. A colleague introduced me to an English producer named Victor, who was familiar with the sport, had traveled extensively around India (meeting his wife at an ashram there), and was also a Kansas City Royals fan on account of his dad being American. On paper, he was the best guide imaginable. Even better, we had a quick chemistry. Within a few weeks, we jetted off to India in the dead of summer. It was sweltering, and we had nineteen days to get our bearings—and, most importantly, to get Sachin to sit down with us.

There was something refreshingly familiar about those hot summer days, something that connected me to my childhood and family. I was communing with something, both my memories and ancestors, and I remembered some of the old stories my mom used to tell my sister and me as a kid. Some of them were about people who went on long journeys to receive advice from a mysterious and wise figure—a long path to enlightenment. On the long journey across the Atlantic, the metaphor became harder and harder to ignore.

Only, those journeys never happen in straight lines. The characters often have limited knowledge and face setback after setback. It didn't take long for us to feel like our trip was going to be doomed.

The first thing we realized: there's nobody in America who is quite as famous as Sachin Tendulkar. On the street, we passed countless shrines dedicated to the player they called "The Little Master," decorated with photos, prayer candles, and flowers. Sachin had another nickname among kids, the quasi-superhero alias the "Master Blaster," because of his supreme power on the pitch and his Clark Kent–like soft-spokenness off of it. We spoke with teammates, coaches, fans, and politicians, and all of them marveled at the cricketer. "Sachin isn't an athlete," Zubin Bharucha, a former Indian National Team player, said. "He's a national treasure. The whole country stops when he's batting."

Paddy Upton, a coach on the 2011 national team, said, "I remember being at one game where a priest, in his full clerical outfit, had a sign that said, 'Sorry, God, I love Tendulkar more than you.'"

Wright Thompson, a senior writer at ESPN who had traveled to India to learn about Sachin himself, explained to me that in many ways, India's relationship to Sachin was akin to a more innocent time in American culture. It was like the 1950s, he said; the country felt like it existed pre-irony, where famous figures could be fully embraced without any worry about who they were dating or what they did out of the spotlight. The best comparison, Thompson said, was really Mickey Mantle—"the Last Hero."

Indeed, nobody we spoke to said anything negative about Sachin. There had never been any scandals. No affairs. Even though he idolized John McEnroe as a child, Sachin was the epitome of class and sportsmanship when he played. "He invested a sense of pride in Indians," said Harsha Bhogle, the cricket commentator. "We can be the best in the world at something." Sachin gave back, too. He helped build schools across the country, educating 160,000 kids.

Slowly, we started to understand this hero's origin story. When Sachin was a kid, he would practice all day, every day. His coach would have to stop him. "It's too dark to practice now," the coach would say. "We're going home." Sachin would drag it out even longer until it was basically pitch black and then, finally, trudge away from the pitch.

That same coach once played a game with him. He placed a single rupee coin—the equivalent of a penny—on the wicket behind Sachin before he batted. The coach said that if anybody could get Sachin out, they'd get the coin. But if Sachin could score a run, cricket's version of a run in baseball, then he got the coin. To this day, even as he's amassed a fortune, Sachin keeps about fifteen of those coins with him. He cherishes them as some of his most prized possessions.

Sachin always set his eyes on the World Cup, we learned. In 1987, he became a ball boy when the World Cup was hosted in India. He strategically made sure that he stood right by the dressing room so that he could study the team's stars up close. He watched how they prepared, how they stayed engaged when not on the field. Sachin learned by example.

All over the city, we heard stories. Everywhere we looked, we saw Sachin's picture. He was a living legend, shrouded in endless lore. But whenever we asked anybody if they knew how we could get in touch, all we received were blank stares.

Sachin seemed to exist just out of reach.

//////

Traveling around India, we met young cricketers who spent hours every day playing the sport, hoping that one day, they'd make the Indian National Team. A few of them were even named Sachin. We met men and women who never cared for the sport otherwise, but found a TV whenever Sachin came up to bat. Rahul Bose, an actor, told us of the 2011 World Cup, "I have never seen such mania." As my team and I learned more and more, we kept hearing one other name mentioned with Sachin. People told us we had to find a way to talk to him. His name was Sudhir Chaudhary. But he's better known as "The Flag Man."

For the past decade, Sudhir had been a staple of Indian National Team cricket matches. At every game, he arrives painted from the waist up—including his entire face—like the Indian flag: green, white, and orange. He'll write TENDULKAR across his back as if it's the name on the back of a jersey, and wave a giant flag. No matter where you're sitting in a stadium, you'll be able to find Sudhir. He covers himself, a type of sacred ritual, with three liters of paint, using two coats so that the colors pop.

We found Sudhir, and he told us his story. Sachin has always been Sudhir's favorite player, and the flag painting was a way to stand out in the crowd. In 2003, Sudhir's friends made a dare. They challenged Sudhir to meet the man that he called his "god." So, Sudhir biked eighteen days—1,200 miles—from his home to Mumbai, where the national team was playing a series against the Australian and New Zealand teams. He discovered the hotel where the team was staying, and as the players passed through the lobby, security tried to push Sudhir and other fans back. But Sachin noticed the superfan; he told Sudhir to come by his house the following day.

Sudhir rode to Sachin's home on his bike first thing in the morning. He waited outside the gate. Soon he was invited inside. Sachin offered him food. "I could hardly eat," Sudhir said. Then Sachin asked, "Sudhir, do you want to watch more matches?" At every national team match since, Sachin has left tickets at will-call for his biggest fan.

Sudhir has traveled the world to attend those contests, always painted in the Indian tricolor. He's been to more than three hundred games since. "I have my family, but today my life is dedicated to cricket," Sudhir once told an Indian newspaper. "My parents tell me to find a full-time job and get married, but I don't listen to them. If I am committed to cricket, how can I find time for marriage and my family?"

Sudhir is sports' answer to a monk. But did he know how we could get in touch with his god? He didn't think he could help us.

We only had two weeks in India. We'd promised Sachin to ESPN. The heat wasn't the only thing making Victor and me sweat.

///////

There are countless other pilgrimages that any sports fan might bike across a country to attend. Sometimes it's about seeing a particular stadium or a particular event. Sometimes it's about seeing a particular player. And if you ever get really lucky, you'll catch all three at once.

We've talked a lot about GOATs thus far, but there are few athletes who have quite the same claim to the title "greatest ever" like Serena Williams. Serena isn't just the greatest tennis player ever. She's the greatest female athlete of all time, and maybe the most dominant competitor we've ever seen—full stop.

In 2016, I signed up to work with Nike to film an ad featuring Serena. We met her, naturally, at a tennis court. This first meeting was really a scout, along with a meet-and-greet to talk about the plans we had for a pretty elaborate commercial shoot sometime in the subsequent few weeks. Before we chatted, I caught her at the end of a workout session, observing her serving, ripping forehands, and smashing backhands. I'd never seen elite tennis quite so close up before, and it wasn't just the power of each shot that surprised me. It was the skill as well. Whether she had time to square up her shot or if she was running off balance, Serena always seemed to place her shot just where she meant it.

At the time, Serena was the number one player in the world—a position she held for a staggering 186 weeks. I wanted to understand not just how she was so dominant but also how she was so successful. In other words, I was curious not just about her power but her skill. In my mind, that would be the substance of the Nike ad we were making. I planned to pitch her on that approach, get her feedback, and then we'd craft a script for her to record when we reconvened on a set in a few weeks to produce the commercial.

So we met in her trailer after her session. Serena sat on a small couch and patiently listened to my pitch. There was no camera with us, but we did have a microphone so we could reference her thoughts and speech patterns to personalize the eventual script. After I was done with my spiel, I waited for some response from her.

Serena didn't have any. She was giving me *nothing*.

I poked a little harder to see if I could engage her to share anything at all. Serena just shrugged and said she thought it was "fine," but it was obvious how disinterested she was in the whole pitch.

I changed tactics. Oftentimes I've used anecdotes of my experiences with Kobe or observations from working with him to break the ice, since so many athletes revere the Black Mamba. I described a particularly intensive workout I'd filmed Kobe doing a few years ago and asked if she related to that experience.

Now Serena stared at me with a look that mixed disbelief with indignation, as if to say *Did this guy really just ask me if I related to working hard?*

Of course she related to Kobe's workout. She trained that hard every single day for as long as she could remember. "Listen, training every day for thirty years is difficult," Serena said—finally. "Then I also think that someone is out there, working really hard, and there's a poster on their wall of me, and they're working to beat me."

Serena spoke almost defiantly: "With that, I get up. I go. Every day I'm sore, but that's the life I chose."

She paused for a moment. "So, there's obviously days I don't feel like training. But there's no day that goes by that I feel like losing."

When we sat down to write the script for the spot, we quickly realized that we wouldn't possibly be able to top her own words. And why would we force Serena to top her own, off-the-cuff delivery? That recording made in a small room during what was supposed to be a pitch meeting was the same audio used for a Nike spot called "Unlimited Serena." It's one of the pieces that I'm most proud of because of how unscripted and even testy Serena is in it. There was no putting words into her mouth or shaping of ideas necessary. Her edge, her revulsion to the idea of losing—there's nothing more Serena than that. We had everything we needed.

Later that year, Serena won her twenty-second Grand Slam tournament—tying Steffi Graf on the all-time list—at Wimbledon. Throughout the tournament, Serena had only dropped a single set. Just as hard to believe, Serena won the singles championship in the morning, and later that afternoon, she joined her sister Venus to win the doubles championship, too.

No day goes by that she feels like losing.

Wimbledon is, of course, one of those traditions that anchor the sports calendar, like the Kentucky Derby in May and the Rose Bowl on New Year's Day. The home of the tournament, the All England Lawn Tennis Club, is one of sports' great cathedrals. It's the type of place built for big moments, a stage to present the best that sports can offer. Every legendary tennis player has played on the famed grass courts since 1877, when the club was founded.

In the summer of 2022, I traveled to London for a few days for work. The trip happened to coincide with Wimbledon. A couple of weeks before, Serena had announced that she would participate in the tournament, even though she hadn't played in nearly an entire year, struggling to return from injuries. I couldn't check the schedule fast enough. Dating back to that Nike ad in 2016, I'd been persistently trying my luck at engaging Serena in a longer project. While I hadn't been successful in wrangling her, I had formed a relationship with her gracious agent, Jill, who always replied to my outreach and told me that eventually the day would come when we'd collaborate further. I reached out to Jill and said I happened to be in London the day of Serena's match, and she kindly invited me and my business partner Ameeth to watch Serena later that week.

It was my first visit to the All England Lawn Tennis Club, about thirty minutes outside London. But it was hardly my first Wimbledon experience. On the contrary, the fabled tournament held a special place in my heart because it was my grandmother's favorite sport and competition. I recall those long, hot summers in India from my childhood, when I would watch Wimbledon with my Nani on a small TV in her bedroom late into the night. That bedroom was the one air-conditioned room in the house—and so we'd sit in there for hours watching Ivan Lendl and Boris Becker and others play. Like I did with the Red Sox, Nani hung on every stroke. She was a tiny lady, but her admiration of the sport was herculean, as were her opinions on each player. Nani had passed away in 2022 at the age of ninety-four, just about a month before Wimbledon began. Because of COVID, I hadn't been able to visit India in the last two years, leaving an ache and emptiness in my heart.

So as Ameeth and I came to the All England club, I brought with me a lot of emotions and longings. The scene felt like walking into an old memory—almost a fantasy. Everything was a perfect shade of green. There were flowers everywhere. I saw the famous Pimm's Cups and strawberries and cream. Some fans paid thousands for the tickets; others camped out overnight and bought them for £30.

When I reached my seat at Centre Court, I had a strange moment of déjà vu. Every year, at the same time, this place dominated my TV for two weeks. It wasn't necessarily a fascination I had with tennis as much a ritual of reconnecting with my grandmother's deep passion and those nights spent in her cool room, watching her watch every shot. I recognized the grass, the lines, the gold-plated Rolex clock. It looked just like the place I knew so well. I had finally made it. Nani would be thrilled.

Serena and her opponent, a French player named Harmony Tan, came out wearing the traditional Wimbledon whites. Serena waved as the crowd roared. At times, Serena looked like someone who hadn't played a tournament in a year. She'd miss a forehand long and shake her head before heading back to try another serve. But then Serena would channel a younger version of herself, and she looked like the player I filmed with years before, seeming to control the ball at will. The game stretched on, going to the final set, and then to a tiebreaker. When she hit winners, the crowd buzzed with a mix of respect and awe and the desire to be a witness to history.

Serena came up just short, only two points from winning the match. We filed out of the stadium and out of the grounds, back to London, marking the end of our pilgrimage. For me, going to the All England Lawn Tennis Club wasn't just a trip to one of sports' most hallowed grounds to witness one of sports' goddesses. It was an emotional, even spiritual odyssey into a deeply entangled web of memories from my childhood, a reconnection with my grandmother, and, in a way, a fitting closure to our relationship, which I hadn't been able to actually have because of the pandemic.

That's another unique power that sports has: It's the fiber that connects us, across continents, and cultures, and generations.

That's what makes it worth the trip.

///////

At the 2011 Cricket World Cup, Sachin and the Indian National Team first faced England. In international sports, there are occasional match-ups that come wrought with history and symbolism. This was one of those situations; the empire vs the colony. The imperialist who first brought the sport to this country centuries ago vs. the place that has made it uniquely its own. The stadium rocked and swayed, and it roared the loudest when Sachin came up to bat.

Sachin started slow, tempered. He knocked a few here and there, feeling the bowler out. He could keep hitting until he got out, requiring a unique mix of power, precision, and patience. Then Sachin saw the first ball he could really drive and *thwack*. He hit the back wall for four. Now he was feeling it. *Thwack. Thwack. Thwack—CRACK.* Sachin hit it off the scoreboard. Fans started to do the wave, and it made it around the stadium five whole times. Sachin kept batting, kept piling up runs. "We Will Rock You" played on the loudspeaker, the whole crowd cheered along, and Sachin sent another one to the fence—bringing his total for the day to over 100, a century as it's called, like scoring 50 points in an NBA game. Sachin finished the day with 120, putting India in a commanding lead.

But after Sachin batted, the Indian team lost its focus. England started chipping away at the lead, slowly. Soon it was down to their final batter, and that feeling of doubt crept into everyone in the stadium. Was India going to blow it again? "It's the most tense stadium I've ever been in in my life," ESPN's Wright Thompson told me. "People are literally praying around me. By hour seven, no one can breathe."

The English player scored. Then scored again. England was down one, with one ball to go. The bowl . . . the hit . . . the ball dribbled right to an Indian player who scooped it up before any further damage could be done. England scored only one run—but that wasn't quite good enough to maintain the Indian victory. The game ended in a tie.

But it was only the group stage. India would play again in a week. The crowds and team would gather again, and Sachin would continue his quest for a World Cup. But now the margin of error was as thin as ever.

///////

My and Victor's quest continued as well—with increasing urgency. Finally, a lead. When we interviewed Rahul Bose, the actor and cricket fan, he had an idea. He wasn't sure how to get in touch with Sachin—but he could get us in to the owner's box of the upcoming game for the Mumbai Indians, the local professional team where Sachin played when not with the national team. Mukesh Ambani, the team's owner and one of the richest men in the world, was close with Sachin. If anybody could get him to speak with us, surely Mr. Ambani could.

We thought the plan was ideal. We expected the box at the Mumbai game to be like an owner's suite in the States, just us and about twelve other people. We'd have plenty of time to plead our case to Ambani. We'd charm him, explain our mission. Then he'd show us the way to Sachin.

I walked up to Brabourne Stadium in Southern Mumbai. It's a huge, perfect circle, built in 1937, when India was still a British colony. It was the first sports-specific stadium ever built in India. At the time, a man known as Lord Brabourne was the governor of Mumbai, then known as Bombay. Brabourne negotiated the purchase of nearly twenty acres of land on which to build the stadium but couldn't come to an agreement on the price. Brabourne offered less; the landowner always countered with more. Just when it seemed like the governor would walk away, he was asked, "Your Excellency, which would you prefer . . . money for your government, or immortality for yourself?"

Brabourne picked immortality—and his name still adorns the sports cathedral built on the expensive land.

Rahul met me and led us inside. "You're only going to have thirty seconds with Mukesh," he said. "Make it count." Thirty seconds? Didn't cricket games last six, seven, eight hours? But when we reached the owner's box, I quickly realized my mistake. This wasn't an intimate setting like stadiums have in America. Five hundred people must have been inside.

We waited and waited, watching the cricket below. Finally, Rahul nudged me. "Here he comes," he said. "Thirty seconds. Do you know what you're going to say?"

Mr. Ambani approached. Rahul introduced us. Chopra. ESPN. Kobe. America. I dropped in as many key words as I could within those precious thirty seconds. Shooter's gotta shoot, right? I asked if he knew how I might get in touch with Sachin. "We've been all over searching for him," I told him.

"I'm hosting a dinner," Mr. Ambani said. "Tomorrow night. David Stern is the guest of honor. And now you as well. I'll invite Sachin. We can see if he'll come—and you can ask him for an interview there."

Wait, what? His key words were as random as mine. David Stern? *NBA David Stern?* I just sort of stared at him. Rahul, who was standing beside me, vigorously nodded and in the absence of any words coming out of my mouth, he replied that we'd be there.

The following morning, I got a call with an address and a time. Later that evening, I got a car to go to Mr. Ambani's house. Only it's not quite a house. It's a skyscraper, twenty-seven floors tall, and covering 400,000 square feet, it's reportedly the second-most-expensive property in the world. It cost over $1 billion to construct. There's a temple, an ice cream parlor, and a fifty-seat movie theater inside. When we drove up to the gates, our driver put the car in park and walked outside to kiss the ground before we entered.

Soon after entering the gardens (modeled after the Hanging Gardens of Babylon), I saw former NBA commissioner Stern chatting away with a group of people while having a cocktail. I waited a few minutes until he was done, then stepped up and introduced myself. My cricket knowledge may have been limited, but my NBA knowledge was deep. Together we were definitely the odd American ducklings in an even odder atmosphere.

Mr. Ambani spotted me and beckoned me over. He was standing with a small group of people.

"The emperor summons," said Commissioner Stern as he pushed me toward Mr. Ambani. "Hit your free throws."

I walked over and stood awkwardly in front of Mr. Ambani. He introduced me to his wife, Tina, beside him. I thanked them for the invitation and started to make some small talk.

He cut me off, uninterested, and then gestured to a gentleman standing on his other side. He was unassuming, 5'5" at best, slight. I hadn't noticed him until that moment. The man held out his hand to shake mine. His touch felt almost gentle.

"I'm Sachin," he said.

Holy shit, I said to myself—and fortunately not out loud. Once again, I could barely stammer a response. I just stared at him. After all the hype and mythology, the odyssey of India, and legends and lore that I'd heard about Sachin, here he was standing right in front of me. I was speechless. It was like meeting the Wizard of Oz himself.

"How can I help you?" he asked.

"Chopra . . . ESPN . . . Kobe . . . America." I stammered my way through. At some point, I asked for an interview.

"Sure," Sachin said, and smiled. In a few days, he'd be staying at a hotel nearby. I could meet him there. "Come over for some tea." He didn't have a ton of time, he said, but he would answer whatever questions I had.

I agreed to the plan and thanked him. He slipped away, and Mr. Ambani likewise said, "Good," and then walked away abruptly.

David Stern slid over. "So?"

"I think I hit the shot," I said.

He grinned. "Well, this will make quite a story one day."

India's next game at the World Cup came against the Irish. They won. Then the Dutch. Another win. Soon they were out of the group stage and into the knockout round.

No matter. The Indian team kept cruising with a win against the reigning champs, Australia. On to the semifinals against . . . Pakistan. The winner would advance to the championship.

When I was a kid spending summers with my cousins, I'd tell them about the Yankees and the Red Sox. I told them it was the most intense rivalry imaginable. About how the two sides seemed to be at war with each other every time they played.

"Sure," a cousin would reply. "But let me tell you about India vs. Pakistan."

Four wars, decades of ill will, nukes pointing at each other . . . yeah, even I could admit that Yankees–Red Sox has nothing on India-Pakistan cricket games.

Sachin, naturally, scored 85 runs in that semifinal, the highest of anyone playing that day. India won. At long last, he made it: Sachin would get one more chance, in his final World Cup, on home soil, to deliver a long-awaited championship for his country. "The feeling of the team was that Sachin Tendulkar had sacrificed pretty much everything in his life to dedicate himself to that cause," said Zubin Bharucha, a former national team player. "The whole team, it seemed was wanting to win it for him."

The final came on a Saturday against Sri Lanka. The whole country froze—and that's not an exaggeration. All day, for miles around the stadium, cars idled in standstill traffic. The team's bus could barely make it to the stadium. The Indian flag was more prominent on that day than it is on Independence Day every August. Horns blared. Fans sang. Everyone watched, and finally, Sachin came up to bat. The cheers came down in the stadium, at bars, and at watch parties from Delhi to Bangalore.

Saaaaaaaaa-chiiin! Sa-chin! Saaaaaaaaaa-chiiin! Sa-chin! Saaaaaaaaaa-chiiin! Sa-chin!

He was out after only eighteen runs.

The country stood silent. It was happening again. They were going to squander another lead, weren't they?

"Tendulkar has carried the burden of his nation for twenty-one years," said Virat Kohli, then a rookie teammate and now one of the world's most famous athletes in his own right. "It was time we carried him." Not unlike Moses journeying to the Promised Land, Sachin got his team there, and then the rest of them had to push the group forward.

The runs poured in. A Sri Lankan player hit a century, but the Indian team kept knocking in runs to counter. Finally, they needed only four runs to clinch the win. Sachin watched off the field, helpless, his fate in his teammate Yuvraj Singh's hands. Singh struck one into the stands. Six runs.

India had won the World Cup at last.

His teammates carried Sachin around the stadium on their shoulders. He waved a flag. Fireworks popped overhead.

At last, Sachin had made it. His journey was complete.

\/\/\/\/

Sachin met us at the hotel right at the time we had arranged. Tea was leisurely. Almost ceremonious in its own way. Then, when we were done, we had a hard twenty minutes to ask questions and film. He was soft-spoken and slight, humble, and in his own unique way, heroic. Never would you imagine that this might be the most famous athlete in the world. He neither looked the part, nor acted like it; if we hadn't spent the previous three weeks learning how revered he was, I'd hardly have picked him out of a crowd.

I asked him about his childhood. "When we won the World Cup [in 1983], I saw it on television and thought, *One day, I want to do the same thing for India,*" he said. Sachin told us about his philanthropy, how he viewed his status in his home country as a "huge honor." Of course, we had to ask about the World Cup and the final. "I was just praying," he admitted.

Right as it was time to wrap up, I asked him how it felt, at last, to finally win—to finally accomplish the goal he had set as a boy. "I don't think anything else can come fifty percent close to that," he said. "It's the only time I've cried on the pitch."

India holds a special place in my heart, not only because it's my ancestral homeland and the place from which my parents immigrated to America. It also has a magic and magnetism to it that I've always recognized in myself. That said, before that journey, I'd never felt Indian. India always remained a foreign land to me until I understood it not just as a physical place, but really as a network of emotions and spiritual fibers that encoded me. When I think of those trips, the odyssey of understanding cricket and tracking down Sachin, as well as the more recent voyage to the All England Lawn Tennis Club to watch Serena play (a fantasy my Nani would only have dreamed about), I realize that those are moments that reconnected me to the mystery of my own existence. It's through those experiences, those inward journeys via the crucible of sports, that we come to really know ourselves.

Isn't that the true embodiment of pilgrimage? It's not just the physical voyage we take across lands and oceans to physical locations and places enshrined with holiness and meaning, but really about the inward journey to those places hidden deeply within our consciousness. Sachin and Serena, a cricket match and Wimbledon—they too are just the crucibles through which we come to fully know ourselves. It's a trip worth taking every time.

With that, we turned off the cameras and packed our equipment. Sachin thanked us for coming. He walked us out to the elevator. "I hope you got what you needed," he said.

We had. I thanked him. He turned to go back inside his room—and then realized he'd left his key inside. He was locked out. The elevator bell rang, and the door opened, jam-packed with folks. I'll never forget the look of the people inside when they saw who was standing in the hallway—a few seconds of stunned silence followed by, all at once, a collection of men, women, and children all screaming simultaneously, all sharing a moment bordering on epiphany.

"SACHIN!"

SACHIN TENDULKAR

NICKNAME: The Little Master

DESCRIPTION: Cricketer

FACT #1:

Considered by many to be the most accomplished cricketer of all time and the greatest Indian athlete ever, thanks to an impressive career that spanned twenty-four years.

FACT #2:

Scored a century—100 runs—in his first-ever professional appearance when he was only fifteen years old.

WISDOM:

"Presence is actually very important in international sport. It is one thing just being there in the middle, but it is another making people aware of your presence. It is about body language and radiating confidence." —*Sachin Tendulkar*

CHAPTER 7

Reformation

Religions don't stay the same forever.
Each generation leaves its mark.

There is nothing permanent except change.

—Heraclitus

MARTIN LUTHER COULDN'T sleep. He felt anxious and nauseous; he suffered from migraines. At the time, he taught theology at the University of Wittenberg, located in eastern Germany between Berlin and Leipzig, although he wouldn't have been there if it weren't for a sign he received from the heavens. Years earlier, Luther was bound to become a lawyer, but riding home from law school on horseback one day, he got caught in a bad storm. Lightning struck just a few feet from him, and the thunderclap erupted with such force that it knocked him to the ground. He cried to St. Anne for help. Somehow he was safe, which he took as a sign. A few weeks later, he entered a monastery and trained to become a monk.

Still, Luther suffered frequently from what he called his *anfechtungen*, meaning, more or less, a crippling crisis of faith. Luther often remembered that moment of panic when lightning nearly struck him as a young man, the feeling of helplessness that resulted. All his life, the church had told him that only the righteous could make it to heaven. Everyone else would suffer for eternity. *Had he done enough?* That question tormented Luther, and when the lightning struck and he thought he was going to die, the answer became startlingly clear: he had not done enough. He felt certain that he was hell-bound. But God gave him a second chance, and now, Luther vowed to not let it go to waste. He dedicated his life to God, but even that didn't allow him to sleep peacefully at night. Even as a member of the clergy, Luther wasn't sure that he was on a path for heaven.

This was the early sixteenth century, and the Catholic Church touched nearly every aspect of life in Western Europe, from schools to governments to charities. The church made it clear that there was always more that parishioners could do to guarantee their spot in heaven. There were masses. Confessions. Communion. Sacraments. Everything had to go through the church directly, a fact cemented by the fact that nearly every copy of the Bible was written in Latin. In many towns, only the Catholic priest could read it.

The church had just begun its costly rebuilding of St. Peter's Basilica in the Vatican, and it had borrowed money from banks all across Europe to pay for construction. Somehow, the church had to raise the money to repay those debts. One solution was to sell letters of indulgence, forms of paper signed by the pope that promised to decrease the amount of time someone would spend in purgatory after death. The more money someone gave, the more time their sins would be forgiven. As one preacher declared, "As soon as a coin in the coffer rings, the soul from purgatory springs."

The practice infuriated Luther. He had altered his life's path to please God, after all. How could the road toward salvation be as simple as handing over some coins? And if things truly *were* that simple, why would God make it so much easier for the rich and powerful to ascend to heaven than those who might not have the money to spend? It didn't seem fair. And it didn't seem like it had any basis in the Bible.

Luther sat and wrote down his thoughts, finishing with ninety-five theses. They were not public disavowals of the church and its policy. This was academic discourse. Each thesis was a question, not a rebuke. "Why does the pope, whose wealth today is greater than the wealth of the richest [man in Rome], build the basilica of St. Peter with the money of poor believers rather than with his own money?" he asked in Thesis 86. Luther did not, as legend says, dramatically nail his document to the door of his local church—at least no historical record supports that version of events. Instead, it seems, he sent it to his archbishop. He simply wanted to start a discussion.

A discussion—that's one way to classify what happened next. Luther's ideas spread throughout the church, which viewed them as dangerous. Meanwhile, those thoughts became more firmly developed in Luther's own head. Where in the Bible did it say anything about the sanctity of the pope? Where did it talk about those letters of indulgence? Luther turned to the Bible for guidance and found one verse that he felt summed up the position he was beginning to take. "The just shall live with faith," it said. His priorities started to rearrange themselves. *Faith* is what mattered. *Faith* is what got people to heaven. Just faith.

Only faith. *Sola fide.*

Church leaders dragged Luther to the city of Worms on the Rhine, where they sought to ban Luther's theses. It was like a trial—with clergymen as the prosecution, and the emperor of the Holy Roman Empire acting as judge. They aimed to give Luther an easy way out of the trouble he had created. They showed him what he had written; all Luther had to do was announce that he no longer agreed with those words. Instead, Luther infamously declared that unless the priests could justify their power and judgment from the Bible, he wouldn't stand down. After all, he said, everybody knew that the pope and the church often made mistakes.

Luther didn't trust the pope? He claimed that the church contradicted itself? That it made errors? The emperor judged that Luther was an outlaw, that his writings were dangerous. If anybody so much as offered Luther a piece of bread, they could be jailed. Luther could be murdered, and the killer would face no legal consequences.

Yet a prince recognized Luther's potential political power, and thus ordered his knights to protect the outlaw. As Luther traveled home, these knights took him to a castle where he could stay in seclusion until, hopefully, tempers had subsided.

At that castle, Luther didn't bemoan his status. He didn't question his righteousness. He continued his humble revolution, continued to serve his Lord in the way that he believed honored Him best.

That's when Luther translated the New Testament into German. He took care to make the translation memorable and accessible. He loved alliteration, wanting people to read the scripture before bed. He commissioned a local artist to create a series of pictures to accompany the text. Thanks to Luther, anybody could follow the Bible and put themselves on the path toward salvation—even if they couldn't read.

///////

Luther's revolution remade the world. Some say that his writing of the theses marked the beginning of modernity itself. He reshaped our ideas of freedom of religion, self-government, and individualism. But one of the most lasting impacts that Luther has on us today is right there in the name of the movement that he helped create: *the Protestant Reformation*. Protest. Then, reform. He helped to entrench that fundamental tenet of change in our societies.

Religions are not static entities. They transform, evolving with each generation. As times change, old practices are interpreted in new ways. Emphasis is placed on different traditions, and traditionalists erupt in anger. In the wake of the argument, while the heart of a religion remains, the ways people apply its lessons shift little by little—and sometimes, all at once.

It's been said that religion is the lens through which we view our world. As a filmmaker, that definition feels especially resonant. Just like I've updated the lenses I use when filming from the ones I used when I first made home movies as a kid, we're constantly updating our faith's lenses as well. As a result, it changes the way we see.

Every religion goes through its own periods of reformation, challenges, and schisms. When Luther first wrote his treatise, nearly all of Western Europe was Catholic. Today, one-eighth of the world's population is Protestant. Muslims famously split into the Sunni and Shiite tribes. After the Buddha reached the state of Nirvana, his followers met for the first Buddhist Council to debate the Buddha's sutras and the budding religion's rules. Buddhist leaders have had several other such meetings throughout history, and Buddhism has evolved in the process. Christianity itself can be seen as the result of a schism with Judaism.

The Religion of Sports isn't immune to this phenomenon. Our beliefs change over time, too, and it's often the young who challenge our preconceived notions. It's people like Kathrine Switzer who showed us that women can run just like men. People like Michael Jordan proved to us that athletes can be brands unto themselves. And it's thanks to Simone Biles that athletes are starting to view themselves as people first. You probably have your own opinions about whether that shift is good or bad—and that's exactly the point. It's exactly how reformation works.

//////

It took only a few moments with her to realize that Simone Biles was different from any other competitor I'd ever been around. Often when I'm beginning a new project, I'll try to get inside an athlete's head. I'll ask them to describe, in minute detail, exactly how it feels to do the thing that has made them so successful. It opens them up, and often they reveal tiny glimpses into their psyche that can help make their extraordinary actions relatable. I figured I'd take a similar approach with Simone. We'd start at the very beginning. "You're standing on the mat about to begin a routine," I began. "What are you thinking in that moment?"

"Half the time, honestly, I feel like I'm just going to die," she said. She explained that the run-up to each vault routine, for example, is like a slow crawl up a roller coaster, the anxious *tickticktick* repeating in her mind. "It's like I'll think of something, and I'll go, 'Oh, I'll never do that.'" But what happens next keeps her coming back for more, over and over. "And then I'll do it, and I'm just like, 'Shoot. This is pretty cool.'"

We often think that everything comes easy for our sports heroes because everything *looks* easy to them. Simone, more than almost any other athlete I've been around, is a master of hiding the difficulty of her work. She seems like a real-life superhero. First, she shields the impossibility of the actual flips and twists, moves so difficult that no other woman has even attempted them. Watching her, you'd never know that what she is about to attempt has never been done. And you *certainly* wouldn't know that she was terrified on the inside. Often, she smiles in the moment that she's most afraid.

Simone possessed no pretense, none of the tough-guy bravado that I'd come to expect from the world's top athletes. She looked me in the eye and, barely knowing me, admitted that she—the greatest gymnast in the world—feels like she's going to die every time she does a routine. Just as surprising, Simone seemed proud of her answer; at the very least, she was totally at peace with it. She wasn't ashamed of her fears. She didn't need me to think she was any tougher than she actually was. She was just herself, without any barriers. The word that comes to mind is *vulnerable*.

I wanted to get a little more specific. My gifted cinematographer, Jessica Young, filmed Simone's practices and caught in beautiful slow motion some of her most acrobatic and gravity-defying moves. We watched it wide-eyed with wonder. "What about the triple double?" I asked her, referencing a skill she first incorporated into her floor routine in 2019. At the time, no other woman had ever landed the move, which is essentially three twists while she flips twice (three twists, two flips . . . triple double). "What does that feel like?"

She thought about it for a moment, seemingly searching for the word. "It feels more like a sound, more than a feeling," she nodded. "It feels like: *zoom!*"

It's one of my favorite answers I've ever received in an interview. It's such a unique response, and it makes no objective sense. A physical, visible act as a sound. Yet when Simone said it, her eyes trailing upward as if she were seeing it in her own head, sounding it out, it absolutely made all the sense in the world.

The net result: even Simone Biles couldn't entirely explain the fleeting nature of her brilliance. That's when it became clear that she operated on her own principles, guided by her own beliefs. As she would show the world during the Tokyo Olympics, Simone Biles is, truly, her own type of superstar.

///////

That first conversation occurred in Simone's hometown of Houston in 2019. She had already cemented herself as the greatest gymnast of all time—but she still had a competitive itch that she couldn't ignore. She set her sights on one more Olympics, in Tokyo in 2020. We were going to film her final few months of preparations and then follow her at the Olympics.

I had come to be well versed in making films about legends in the twilight of their careers, having spent time with Tom Brady, David Ortiz, and Kobe Bryant long after they had proven themselves. In many ways, we envisioned this project with Simone as following a similar template, with one major caveat. Simone was also one of the youngest athletes I had ever covered. When we first met her, she was just twenty-two. Can you imagine how strange it must be to talk about the end of something, right when your adult life is beginning?

But that's just the way gymnastics works. Unlike almost every other sport, the life cycle of a gymnast's career is remarkably abbreviated. Because flexibility is so critical to success, and because the constant harsh landings can punish a body, many gymnasts don't compete after their twentieth birthday.

There's simply no other competition on the gymnastic calendar that compares to the pedigree of the Olympics—and because a gymnast's window of peak performance is so small, most competitors have only one chance to qualify and compete. Often, the second a young gymnast shows any promise (even if they are only ten years old), a date six or eight years in the future is circled on their calendar. That's *their* Olympics. That's their one shot. That's what they'll be working toward for nearly the next decade.

Simone got involved in gymnastics by accident. When she was six, her class was supposed to go on a field trip to visit a ranch outside Houston. But it rained that day, and the teachers had to scrap the outdoor excursion. Instead they changed plans and took the group to a gym called Bannon's Gymnastix. Some of the coaches there noticed how one particular kid in the class watched older, more experienced gymnasts and mimicked their moves almost perfectly without any practice. When Simone went home that day, she brought along a note from the coaches. It suggested that she sign up for more classes. Her parents signed her up.

Watching the Biles family's home videos of Simone from this time makes it clear that her talent was all but obvious from the start. She's tiny—not even five feet tall—but tumbles across the floor and floats between the uneven bars like a master. She started to win tournaments. Once, when a crowd of college coaches came to scout her, Simone's father asked one of them, "Coach, why are you guys looking at Simone? I don't understand."

The coach couldn't believe the comment. "Have you seen your daughter?" he said.

"I see her every day," Ron Biles Sr. said.

"Well," the coach replied, "she's a freak."

"I beg your pardon!" said her father.

The coach tried again. "I mean that in a nice way."

Like that, Simone was on the path to the Olympics. And for someone with her combination of talent and drive, success followed soon thereafter. Team USA invited her to train with the national team, and in 2016, she qualified for the Rio Games—the ones that had been circled on her calendar for a decade. You probably remember how she became a sensation that summer, winning five medals: four gold and one bronze. Before her twentieth birthday, folks talked about her with reverence. Yet while the rest of the world gawked at her ability, she was in awe at the moment itself, taking to Twitter, Snapchat, and Instagram to share inside looks at what it was actually like to be an Olympian. Most of the time, those captions were summed up with three letters: "W.O.W." After those Olympics, Simone was anointed as the greatest of all time—and to wink at that fact, she started to wear leotards to competitions bedazzled with the head of a goat. You've got to admire the swagger.

Most gymnasts' careers would have been over then and there,

considered a whopping success. But Simone wanted to prove that she could do it again. She announced that she would try for *another* Olympics in 2020 and then would call it a career at twenty-three years old. The experience of being an Olympian was so powerful, so hard to believe, that she just had to live through it one more time.

That's where my team started documenting her path for a series called *Simone vs Herself*. We started filming in the latter half of 2019, building toward the following summer, when Simone would be off for Tokyo. Everything was planned. In fact, every one of Simone's practices for the previous four years had been plotted to maximize her potential in August 2020, when the Olympics began in Tokyo. Practice. Rest. Diet. Repeat. She always had a singular focus.

In January 2020, far away from the world of sports, a public health issue started to pop up in the news. As we traveled back and forth from Los Angeles to Houston to film Simone's training, nobody worried much about the coronavirus. But over the coming weeks, the disease spread, while concern grew. Before we knew it, the world came to a standstill.

Simone was practicing her floor routine when she got the news that the Olympics had been postponed due to the global pandemic. One minute she flipped and twisted, and then, in the middle of a break, she checked her phone. Texts flooded in. "Have you heard the news?" And just like that, her world was thrown completely off its expected course. She sat in the locker room and cried. Our production team FaceTimed with her soon after the announcement. "If they cancel the Olympics, then I'm quitting," Simone admitted.

Without the Olympics to strive for, what was the point?

/////////

When lockdown began, Simone barely left her house, just like the rest of us. Her coaches sent her workouts to do at home, which she supplemented with long walks around the neighborhood with her dogs Lilo and Rambo. She tried a couple of hobbies, but none of them gave her the same joy as gymnastics. "During that quarantine process, I got to really relish in my emotions for the first time in my life without any outside voices telling me that it was going to be okay, or I can do it," she said.

It's not as if Simone had never faced challenges off the mat. As a young child, her birth mother struggled with addiction, and some nights, Simone went to bed hungry. When they were toddlers, Simone and her three siblings were placed into foster care. Some nights in the foster home, Simone ran to her brother's room and held on to him tight, worried that he might disappear. When their extended family heard that they were in foster care, they hatched a plan. The two eldest siblings would live with their aunt and uncle in Ohio; Simone and her sister would head to their grandparents in Houston. Eventually, those grandparents became "Mom" and "Dad" once they completed the paperwork necessary to make the adoption official.

Simone loves her adopted family and recalls those early days with fondness. As a kid, she would grab her adopted brother Ron's arm and do pull-ups. "This is not normal, man," Ron remembers thinking at the time. "This is crazy." In home videos, you'll see a young Simone in bright pink pajama pants tumbling through the dining room, three backflips in a row punctuated with a stylish finish. A perfect 10.

When you're around the Biles family today, all you hear is laughter. They grew close, Simone kept dominating the gymnastics world, and soon she got invited to join Team USA at the legendary Karolyi Ranch, about an hour away from the Bileses' home in Texas. Everything seemed to be going exactly according to plan. Once she trained with Team USA, she'd have a shot at the Olympics. Simone couldn't wait to go to the ranch.

The Karolyi Ranch became the home of the American team's training camps in 2001 and since then had earned a reputation for the rigid rules that coaches required athletes to follow while there. "I thought it was summer camp, and it was not summer camp," Simone remembers. "I like to have fun when I do gymnastics. That was not fun." But to get to the Olympics, she had to follow the Karolyi system, which meant keeping her head down and never getting the satisfaction of a coach telling her, "Good job." At the ranch, there were five stations: vault, bar, beam, floor, and finally, therapy, where the gymnasts met with the team doctor, Larry Nassar. In 2018, Nassar was sentenced to more than one hundred years in prison after pleading guilty to multiple sexual assault charges brought by Team USA gymnasts.

"All those years, nobody ever told us what sexual abuse was, so we didn't really feel like we were going through it, or we were victims," said Simone. "A lot of us didn't go to school. We were homeschooled, so it's not like we had people to talk about it with." But one time, years later, Simone approached one of her friends. "Hey," she asked. "If I've been touched here, have I been sexually assaulted?"

Simone thought she was being dramatic, but her friend knew otherwise. "Absolutely," the friend said.

"Are you sure?" Simone replied.

Even then, Simone wasn't sure how to process Nasser's abuse. Simone explained that she remembered leaving that conversation thinking, "*I don't think so*, because I feel like in those instances, I was one of the luckier ones, because I didn't get it as bad as some of the other girls I knew."

When the news about Nasser's abuse came out following the Rio Olympics, her mother tried to broach the subject. Simone's reply was always curt. "No. Don't talk to me. That didn't happen to me," she would say.

One day, she was driving on Highway 99 in Houston when the

flood of memories hit her all at once. "I just remember calling my mom," Simone says. "She told me to pull over."

"Can you drive?" her mom asked.

"She didn't say anything," her mom remembers. "She just cried, and we cried together because I knew what it was that she wanted to talk about. She didn't have to say anything."

Simone didn't leave her room for days. She slept a lot. "With gymnasts, if you get injured, you're like, 'Okay. Your heal time is four to six weeks.' But then with something so traumatic that happens like this, well it's not four to six weeks, so it's hard for us to process that. There's no time limit or healing time for it, so you just take it day by day."

From the inception of the series, Simone wanted to include her story as a survivor of sexual assault. She knows that her voice might help somebody else who is struggling. "It's okay to say, 'I need help,'" she said. "There's nothing wrong with that."

Watching Simone Biles compete, you would never know that all these complicated thoughts and trauma help fuel her. To me, Simone is a master of resilience as much as she is the best gymnast in the world. We watched her practice for hours, and not every landing is perfect. She falls a lot, but always, she pops up and tries it again.

For so long, in the face of so much hardship, that had been Simone's coping method: *go go go.* Always compete. Always go for gold. Always get back up. Always forward.

Now Simone *couldn't* charge forward, even if she tried. She had to confront the demons at hand head-on, and that contemplation made Simone reaffirm her goal: she swore that she would make it back to Tokyo. "Deciding to still go forward and train for the 2021 Olympics was only up to me, no outside voices," she said. "That's what I was fighting for and training for."

Finally, the gym opened again, only for her and her coaches. She felt ready physically. "Mentally, on the other hand," she said, "I was really worried because I'm getting older, getting tired. It's getting harder to get up and go to the gym." Normally, Simone was the type to arrive at practice early and stay late. Now she would show up just in the nick of time before her training began.

Still, she made it to every session. And there were a lot of them. To get her body ready for the Olympics, Simone practiced six times a week, twice a day. She only rested on Sundays. "Having to come back, you have to bring a whole new level," she said in the months before the Games. "Then I worry about if I'll be just as good. That's my big fear . . . to see if I can get back to 2019 shape and ability. Can I do it again?"

A few months later, when Simone's gym fully reopened to the public, she would practice at the same time as young girls who were just beginning their training. One of my favorite things to do was to watch their reactions after Simone did a drill. Everything else in the gym stopped whenever Simone was working. Every head turned. Watching those girls watch Simone, I would see some of them look astonished, others a little intimidated. After a few particularly improbable moves, the gallery would break out into applause. Gymnastics is the only sport I know where the distance between the youngest competitors and the greatest of all time is so razor thin. Imagine a gym where on one side you have a ten-year-old basketball player learning how to dribble, and on the other you have Steph Curry working on his step-back three. Then imagine Steph, while catching his breath, walking over to give the slack-jawed ten-year-old some pointers. It doesn't happen—but it does with Simone; her genius is right there. Aspiring gymnasts could touch it, feel it, interact with it. And they could because their heroine engaged with and shared her gifts in a way that no other legend ever could or would.

Sometimes, you'd see a young girl internalize one of Simone's moves, and then slowly try to twist their body in the same motion. That girl would then whisper something to one of their friends, and I always liked to imagine what they were saying. I thought it was probably something like this: "One day, I want to go to the Olympics, too."

////////

We planned to send a crew to follow Simone to Tokyo, to check in with her every few days, to follow her during training and throughout the competition. But once again, things didn't quite go to plan. With COVID cases spiking, Olympic officials restricted access to the Games. Even Simone's own family couldn't attend her competitions or visit her in the Olympic Village.

Instead, we sent some equipment with Simone and scheduled times for her to turn the cameras on while we interviewed her via Zoom. Add one more talent to Simone's résumé: in addition to being an activist and an Olympic medalist, she's also an adept cinematographer.

The day before Simone's first competition, she turned on the cameras. Simone sat in her room wearing a black hoodie with a goat on it, a nod to her status in the sport. She smiled and seemed excited. She said all the right things. "We're here," she told us. "We made it. Whatever happens happens."

"If I'm being completely honest," she said, "I'm feeling ready."

Through the computer screen, we asked her how the approach to this competition might be different than others. "This time of year, you've done so many routines, done so many repetitions, now you just kind of go on airplane mode," she said. "Even if I'm feeling a little bit out of it, I know I can go up there and hit a set. And that's the beauty of training so many routines over and over and over again." She paused for just a second, looking off into the distance. Something seemed to shift. She furrowed her brow, and I couldn't tell if what she said next was a statement or a question. "Even if you're not a hundred and five percent ready, you're still a hundred percent ready—so you're ready enough."

I didn't notice it at the time, didn't detect those first signs of doubt in her voice. In retrospect, knowing the turn her story took, I should have realized that something might not be right. Instead, I just thought it was typical athlete-speak, and the wavering in her voice was only the result of her looking ahead, focusing on her qualifiers in the morning. I didn't realize that she was about to change the very way that athletes everywhere speak out.

Qualifying for the team event went poorly. Instead of the perfect-10 landings we'd become accustomed to, Simone stumbled on nearly every one. The Americans trailed the Russians in points. When she logged on for our interview that night, she didn't have to speak for her frustration to be evident.

"I feel like I just put too much pressure on myself," she said. "I just feel like mentally I'm struggling. I just feel like I'm overthinking everything."

Then she stood up on her bed. Simone was frustrated now. "Here I am in my frickin' hotel room doing standing flips on my bed like an idiot," she said. Then she jumped on the bed, did a backflip, and wavered on the landing trying to catch her balance. "Like, I *know* how to do gymnastics," she said, talking to herself as much as anyone. Simone jumped again, did another flip, but couldn't get her legs under her. She landed on her back.

The greatest gymnast of all time was left jumping on her bed. She had a case of what is known as the "twisties," a phenomenon that causes a gymnast to feel lost in the air, not unlike a case of the yips in golf. The team finals started the next morning. If she couldn't stick the landing on her bed, how was she going to do it in competition?

Our team watched the live coverage on TV in Houston with the rest of the Biles family. Her brother said a prayer beforehand. Soon it was time for Simone to do the vault. As she prepared to go, I couldn't stop thinking about what she had told me the first time we'd met.

Every time she did this trick, she had said, she thought she might die. Surely, I thought, this would be yet another superhero moment where she conquered that fear, conquered the laws of gravity, and zipped through the air in a way that surprised everyone, including herself—in that way that's best quantified by the sound she hears while doing it. When Simone began her skill, though, she didn't do the two-and-a-half she had been practicing.

As I said, I often find that one of the best ways to connect with athletes is to ask them to walk me through their performance step by step. When my senior producer and main creative collaborator, Katie Walsh, spoke with her later, Katie asked Simone to walk her through this performance, too. "I had no idea what I was about to do," Simone said. "But I knew I wasn't going to do a two-and-a-half, because I could just feel it in my body. I didn't know how to pull a twist. So I was just running and blocking and praying . . . I had no idea where I was." She added, "I just felt like a flying rag doll."

On TV that day, we could see Simone talking to her teammates. "I don't want to do something stupid," she said to them. Then she walked off into the dressing room. Nobody at the Bileses' watch party in Houston was quite sure what was happening. And then—"Mom! Mom! Mom!" said Simone's brother Ron Jr., holding a phone. "It's Simone. Talk to her."

Everyone in the room—brothers, sisters, cousins, friends, and our crew—listened to one side of the phone call.

"You can't do it?" said Simone's mother, Nellie.

Then, "That's okay."

"That's okay, honey."

"They will do their best without you. I don't want you going out there if you're not in a good place."

"You don't need to go out there and hurt yourself. That's just not right."

"You need to take care of yourself."

"Okay?"

"Okay."

"I love you. Just take some deep breaths and just know we're praying for you, okay?"

"Okay, well, good. Well, hopefully after this you'll feel better."

Nellie turned to address the rest of the group gathered at the watch party. "That's it," she reported. "She's done. She just pulled out. She says she can't do it."

///////

The news spread out of that room not long after that. *Simone Biles has withdrawn from the team finals.* Shortly afterwards, speculation started that the problem wasn't physical for Simone; it had something to do with her mental health. Almost immediately, Simone became a flashpoint for the culture wars. One side hailed her as a champion and a trailblazer. The other thought she represented everything wrong with young people today, calling her entitled, selfish, and weak.

Simone sensed the blowback to her decision as well, and it was clear—in those first few days following her decision—that it stung her that people didn't see things her way. "It's like, people have absolutely no idea what the twisties is, and they think, 'Oh, she quit. She lost a medal for the country,'" she told her family over FaceTime. "I'm like, 'Actually, first, I didn't lose any medal. If I would've continued, I probably would have gotten injured. They wouldn't have been able to replace [my scores]. They don't even know the process of it. They just think I quit. I don't think they realize that on the vault, I didn't do what I was supposed to. They didn't even process it."

As we've already established, sports have myths. And one of those is that no matter what might be ailing them, great athletes never quit. We expect them to play through pain, to grin and bear it. Like our old friend Pheidippides, the Greek messenger, athletes are hailed because they'll run until they literally cannot run any farther. Some of our great sports heroes are remembered because they conquered mental and physical pain: Cal Ripken Jr. and Lou Gehrig refusing to come out of the lineup, quarterback Brett Favre throwing for nearly 400 yards on *Monday Night Football* the day after his father died, and Michael Jordan playing with the flu, to name a few examples.

When Simone Biles dropped out of the Olympics in Tokyo, she wasn't just bowing out of a competition. She was turning her back on one of the most foundational beliefs in sports. She was showing that she interpreted the faith differently than we'd become accustomed to. She said she could be great *and* look out for herself.

Simone Biles led a reformation of the Religion of Sports.

After the Olympics ended, Katie asked Simone how she would sum up her career to that point. "Courage," she said. "Standing up for yourself. Resilience. Bravery."

///////

Revolutions don't happen overnight. They are movements that start slowly at first, then grow and grow until they burst clearly into view in one instance. Simone might have provided a highly visible moment when the world first took notice of the value that athletes place on mental health. She even might have been among the most vocal about it. But she certainly wasn't the first to recognize just how important being dialed in mentally is to performance. This is something I've become acutely aware of working with various athletes across my career.

Ask any football fan, and they will tell you that Michael Strahan is among the most feared defenders in NFL history. In a macho sport, he's as tough as they come, a huge defensive end who holds the NFL's single-season sack record. I had gotten to know Michael after his playing career as he became a mogul off the field in the entertainment world. When I started to conceptualize the idea of Religion of Sports as a company, I wanted Michael involved not only for his sports credentials but also because I valued his keen eye in the business world. He's one of the most self-aware and brilliant people I've ever met.

Of course, whenever I see Michael on a Zoom or at a board meeting, we always wind up talking at least a little bit about sports (and a lot about his Giants meeting my Patriots in 2008 to spoil the Pats' perfect season). However, one time, we were discussing a player (I'm not gonna name names, but I think you can guess) and whether he would retire after the next season, and I speculated that his body couldn't handle the load of another seventeen-game slog. Michael wouldn't hear it. He said that for the great athletes, it's not the physical wear and tear that undoes them. They're professionals, after all, and a decade into their career, most athletes know how to take care of themselves. They know what they need to do during the off-season as far as training, nutrition, and recovery go to get their bodies right for the next season. "It's the mental and emotional part that's no longer tolerable," Strahan explained. He said it happened to him after Super Bowl XLII. And it would have happened earlier, he said, if it weren't for an unlikely meeting that changed his career.

Michael shared the story. It was 1998, and Michael was well on his way to his second straight Pro Bowl and a 15-sack season, the third-highest total in the league. "But I just didn't feel it," Michael said. "I just didn't feel like I was doing everything that I could do. I wasn't as good as a player as I expected myself to be—and I couldn't figure out why."

Week in and week out, Michael performed. But he didn't feel like himself. "I just felt—mentally and physically—a step behind," he said. "It was almost paralyzing." In week 13 Michael faced Steve Young's San Francisco 49ers on *Monday Night Football*. By every account, Michael provided one of the Giants' lone bright spots in a humiliating 31–7 loss. He had two sacks, and when he came to the sideline, his teammates and coaches patted him on the back. He waved his arms and roared, trying to signal that he was pumped up, dialed in. But that's not how he actually felt. Not at all. Michael felt lost. He felt like he was playing horribly. "You almost feel like a farce," he said. "If people looked, they would think nothing was happening. They'd think, *Oh, he's doing great.* But I just know I'm not. Like, something is *wrong.*"

Michael kept those thoughts to himself. "I didn't confide in any teammates. I didn't confide in any coaches. I didn't want anybody to see that," he said. "We're taught to be macho. We're taught to figure it out yourself . . . '*When you got a job to do, I don't care how the job gets done, just do it.*' That's what I was doing. I was trying to posture . . . to fake it till you make it. And I just couldn't fake it anymore."

Michael belongs to a generation—and until very recently, a sport—that is "old school" to the extreme. Football players—and certainly coaches—didn't believe in the idea of anything but meeting a challenge head-on. "Mental toughness" was a characteristic prized above all else; "mental health," on the other hand, had been a perplexing idea. If anything, this had been the attitude that pervaded sports—not just football, but all of sports—and for decades, it made up the ethos of the Religion of Sports as well. So for Michael, whatever solutions he was going to look for to solve his mental slump, he was going to do it discreetly.

The next week, the Giants faced the Arizona Cardinals, but instead of flying cross-country twice in a few days, they decided to spend the week in Tucson, Arizona, and practice at the University of Arizona's facilities. Sitting on the college campus, Michael decided to get some help. He called a sports psychologist who lived nearby. The doctor told Michael to come over, and so one of the NFL's meanest men drove out to the desert to see a shrink.

"It was the first time I'd ever seen somebody like this," Michael explained. He told the doctor about how he'd been feeling, about how he felt like a fraud. They talked for hours, and finally, the psychologist suggested an exercise. He told Michael to lie down, to breathe. He told him that every time he inhaled, he should imagine clean air coming in, and every breath out would release the dirty air leaving his body and mind. "As I do it, it's like a gas meter. I just picture the good air coming in, and the bad air coming out all the way down my body," Michael said.

Without describing it as such, this is a form of mindfulness—or if you want to get even more far out, meditation. Observing one's breath and becoming self-aware of its regulation in and out of the lungs is a traditional technique that goes back millennia in Eastern wisdom traditions. Michael's New Age teacher didn't bother to share that context with him. Michael admits now that if he had been told all that then, he may not have felt comfortable doing the technique back then. But he still does the breathing technique to this day. He did it before every game for the rest of his career.

But the first test was that week against the Cardinals. How did it go? "It was truly as if I knew what they were going to do before they knew what they were going to do themselves," Michael said. "I'd never been that focused and that confident and that sure of myself. I'm in the backfield, and I'm making plays. I'm getting up, like, 'Where did that come from?' I felt like Superman."

He continued, "From that one experience, it completely shaped the next ten years of my career. The ability for me to switch my mind to literally internalize and talk to myself and switch it, I never went into a game not feeling confident, not feeling like I was the best guy out there."

Today's athletes build on the example that people like Michael helped set. Only now, stars like Naomi Osaka and Kevin Love don't feel like they have to hide their mental health struggles. Thanks to them, Simone, and others, athletes today feel more emboldened than ever to prioritize their mental well-being over the constant pressure for a championship. "By being open about your struggles, you flip vulnerabilities into victories," wrote Love on Instagram after Simone withdrew.

Another example is Mikaela Shiffrin. The star skier was expected to contend for five gold medals at the Beijing Olympics just a few months after Simone's run in Tokyo. On the outside, the world saw the best skier in the country. But internally, Shiffrin struggled to cope with the sudden death of her father earlier in the year and a slew of recent injuries. As she tried to compete, her Olympics looked eerily similar to Simone's.

Shiffrin entered one race and skied out of bounds, something she almost never did. Then she did it again. For about twenty minutes, Shiffrin sat on the side of the mountain right in the snow, trying to figure out what to do—and what was wrong. As she sat there, a simple thought came into her mind. Shiffrin explained to ESPN, "There's a scene in *Frozen II* where Anna sings, 'Just do the next right thing.' That was the thought. Not, 'What is the next day going to bring?' Or, 'What is the next race going to bring?' Just, 'What's the next step I have to take?'"

In the months following the Olympics, Shiffrin discovered those steps. First, she started speaking out on the importance of mental health—and she says that in no small way did Simone's actions in Tokyo inspire her. "That was a turning point for a lot of athletes, including myself," Shiffrin told ESPN. "We had talked about pressure and mental health—maybe we didn't call it that—before, but I felt if I talked to the media about the anxiety I felt, they were like, 'You're just not cut out for this,' and I felt like a bad person for feeling that way and succumbing to the pressure."

Shiffrin also did something else. She kept skiing. It didn't take long for her to rediscover her focus. Just a few weeks after the Olympics, Shiffrin competed in the World Cup finals in Courchevel, France. She won the title, and with it, a gigantic glass trophy made to look like a globe. "I really don't know if we actually have a plan to get that thing home," she told the *Washington Post* after her race.

"It doesn't count as carry-on luggage, but you definitely don't want to check it," she continued. "It gets a little complicated. But that's okay."

A little complicated. But that's okay. Kind of like the way Biles and her acolytes have shifted our understanding of the world's top performers.

/////////

Let's make one thing clear: nobody can doubt whether Simone Biles is tough. She had done the grin-and-bear-it exercise before and performed admirably. "I've done gymnastics on broken ribs, two broken big toes—or shattered, because they were not just broken. They were shattered in pieces," she told me after getting back from Tokyo. "Kidney stones. I've been through sexual abuse. I came back to the sport. There are so many barriers that I've gotten past and to say that I just had a bad turn and quit, if you look at all of those, you can see I'm not a quitter. I'm a fighter."

If you need any more proof of her resilience, you just have to look at what happened next in Tokyo.

All day, Simone went to the gym to practice. She'd flip and land on her back. On her knees. Everywhere but her feet. She tried everything to get herself feeling like the greatest ever, but nothing worked. Simone would flip, she'd fall, and she'd scream in frustration when she hit the padded floor.

Right before the beam finals, the final chance for Simone to compete, we interviewed her again with our remote setup. "Part of me was feeling confused, anxious, scared," she said of her mindset during practice. "And then it turned into almost annoyed, mad. And then it seeped into, 'Okay, I'm okay with this. This is supposed to happen.'"

Was she feeling ready? Not quite. But was she going to compete? Absolutely.

"I'm doing this for me," she told us. "I don't care if I fall five times. But I'm getting up, and I'm going to compete one more time because this is my dream."

Simone adjusted her routine. She was going to perform a double pike coming off the vault, a relatively simple move she hadn't done since she was twelve or thirteen. She stood at the edge of the mat. She ran. She vaulted. And she stuck the landing. Not her best vault ever, but a huge smile spread across her face. At home in Houston, her family cheered and hugged. Teammates cried. She won bronze.

A few weeks later, when she returned to the States, Katie, Jess, and the team met with Simone in New York. She showed us her bronze medal. "This bronze feels like a gold to me," she said with a grin. "I don't care what y'all say."

Simone is at peace with what happened in Tokyo. "I didn't know who I was, but now I feel like it's surfacing," Simone confessed. "It's unfortunate, because it took the Olympics for me to realize that, but I'm so happy I did. Because now I know I'm worth more than gold. Rather than if I won, I'd walk away from the sport only being known to win, and there's kind of a certain type of way I would have felt about it. At the end of the day, people [now] respect me as a human, as an athlete: Simone. That's what I've always tried to get people to see."

In that quote, Simone seems to give her answer to one of the questions that her actions posed. Like Luther, she points out the difference between her interpretation of her religion and the one that came before her. Simone believes that she can be great, *and* she can look out for herself. That she can be a competitor *and* be compassionate. That life is about more than just results. That she isn't measured just by victories. That grit doesn't have to be dangerous.

Even if the way she practices it might not be familiar to you, Simone still achieved a title that any sports fan recognizes: she is—in every sense of the word—a champion.

SIMONE BILES

NICKNAME: The GOAT

DESCRIPTION: Gymnast

FACT #1:

Widely considered the best and most dominant gymnast of all time, Biles has won 5 Olympic medals and 25 World Championship medals.

FACT #2:

Her impact on her sport is so complete that there are *four* different gymnastics elements known as "The Biles": one on the beam, one on the vault, and two on the floor.

WISDOM:

"I'm not the next Usain Bolt or Michael Phelps. I'm the first Simone Biles." —*Simone Biles*

CHAPTER 8

Tribes

Time to make one thing clear: it's not the Religion of the Red Sox and the Religion of the Yankees. We all believe in the same thing. We're just part of different tribes.

The basis of world peace is the teaching which runs through almost all the great religions of the world. "Love thy neighbor as thyself." Christ, some of the other great Jewish teachers, Buddha, all preached it. Their followers forgot it.

—Eleanor Roosevelt

NOBODY EXPECTED THE war to last so long. When World War I began in late July 1914, people assumed the fighting would be over in six weeks. Parents encouraged their boys to hurry to the front; if they didn't make it soon, they'd miss the excitement. *All our boys will be home for Christmas,* they thought. But as the two sides dug in—literally—it became clear that this war would be a slog. It would also be grueling. Men suffered from trench foot. If enemy fire gunned down a soldier in no-man's-land—the area between trenches that could be as little as fifty yards—the body could lie there for weeks, rats gnawing the flesh to the bone, since retrieving the corpse was far too dangerous. The trenches eventually twisted from the border of Switzerland all the way to the North Sea.

The scene by Christmastime 1914 was bleak. French casualties already numbered 306,000, the Germans lost 241,000, and the British lost 30,000. One of the war's first major battles, the First Battle of Ypres, dragged on from October to November, only to end with both sides giving up.

Back home, each side painted its opponent as evil and cruel, and studying the First World War, you can see why many people tend to view tribalism as a lamentable human impulse. As the famed Russian-American writer Ayn Rand once remarked in a speech, "There is no surer way to infect mankind with hatred—brute, blind, virulent hatred—than by splitting it into ethnic groups or tribes." War makes us pick sides, and in those first months of conflict, the British made posters characterizing Germans as bloodthirsty monkeys, sucking the life out of the earth. In Germany, the British became imperialist monsters; a spider whose web stretched all over the world.

But the characterization that tribalism is always problematic is oversimplified; there's nothing inherently wrong with tribes. Separating oneself into groups is a primal instinct that goes back to the earliest days as hunter-gatherers, and to go against that instinct is unrealistic. Tribalism goes wrong only when those instincts get channeled toward exclusion, where the unique power of "us" shifts toward the poison of searching for the "other."

There are different, healthier ways to split into groups. The best way to funnel those tribal tendencies into something greater is through sports.

Let me tell you a story.

///////

The situation on the front lines in Belgium and France grew more and more dire with each passing day in December 1914. For weeks, rain came in sheets. The trenches flooded, men bailing water over the sides and quickly ducking their hands back down so as not to draw machine-gun fire. Pope Benedict XV viewed it as his duty to encourage peace. In early December, he begged leaders from both sides of the fighting to end hostilities for good. They were all Christians; they all worshipped the same god. Couldn't they set aside their differences?

Neither side agreed. When that didn't work, the pope wrote a letter. Please, he prayed. At the very least, let it be "that the guns may fall silent at least upon the night the angels sang." Surely, neither side would slaughter the other on Christmas.

But military leaders wouldn't hear it. War didn't take a holiday. Any Christmas celebrations on the front lines would be subdued and colored with the threat of more gunfire. Each British soldier was rationed a chocolate bar and some tobacco. Merry Christmas, from the king. Don't let your gun leave your sight.

On Christmas Eve, a Thursday, the temperature dropped, and the unrelenting rain turned into a light snow. The muddy no-man's-land was blanketed in white, and when night fell on the British side, some men tried to go to sleep while others stood watch, waiting to warn the others of enemy fire.

The British troops scanned the top of the German trenches for any movement. Suddenly came a dim yellow light. Then another. Then another. What could it be? It was too faint to be searchlights, and it made no sound like gunfire. As more and more of the lights appeared, their form became clear. They were candles, and they illuminated another decoration: a series of Christmas trees that decorated areas of the parapet typically reserved for artillery. The German government had sent the decorations to boost morale and, hopefully, keep their soldiers dedicated to the cause. But the Germans didn't want to fight now. They started to sing "Silent Night," and indeed, the night did grow silent then. The British listened, and when the Germans finished, the British began to sing "The First Noel." It continued like this for some time; the Germans sang, then the British. Eventually, while the British sang "O Come, All Ye Faithful," the Germans joined in. "I thought, *Well, this is really a most extraordinary thing*," remembered one English rifleman named Graham Williams. "Two nations, singing the same carol, in the middle of a war."

The caroling continued throughout the night, and letters written home reported that barely a soldier slept. When daylight came, at one point along the trench, a German held up a sheet of paper with a handwritten message: "You no shoot, we no shoot." When they heard a cheer in response, they crept over the top of the trenches. It was a move that never could have happened under normal circumstances. But the British soon crept over, too. The two enemies walked toward each other, and a little uneasily, they shook hands and wished each other a Merry Christmas.

Here they were, two tribes literally at war, bonding over a common cause. Two sides coming together thanks to their shared religion. "The Christmas Truce," as it has come to be known, often gets held up as proof of the power of Christianity. Sure, the men all celebrated a holy Christian holiday. But what did they actually do when they dropped their weapons and met on the battlefield that Christmas Day?

They didn't read from the Bible. They said no prayers. No soldier delivered a poignant sermon.

They played soccer.

Somebody, somewhere, had a ball, and they threw it into the middle of the celebration. Germans kicked the ball to the same British soldiers whom, just hours before, they had been trying to kill. At some places on the trench line, soldiers participated in giant games of kick about, sometimes including hundreds of men. When they didn't have a soccer ball, they fashioned one together from scraps of old uniforms. At other spots, more organized games occurred.

They took off their caps and used them to mark the goals. One such game featured the Royal Army Medical Corps taking on the 133rd Saxons. The Germans beat the English, 3–2. According to one British lance corporal's account, a German approached him during the day speaking in broken English. He said, "I would like to see Arsenal play Tottenham tomorrow."

The whole scene—warring sides playing a game of soccer—is so fantastical that it is frequently dismissed as urban legend, but enough contemporary accounts of the festivities exist that it's fair to say that the games almost certainly happened. In 2014, a statue commemorating the centennial of the event was unveiled outside of St. Luke's Church in Liverpool. It depicts a German and British soldier extending their arms to shake hands over a soccer ball. Again, this is a statue about Christmas that was erected *outside of a church*—and the importance of soccer is placed front and center. "A lively game began," wrote German lieutenant Kurt Zehmisch in his diary on Christmas night, 1914. "How fantastically wonderful and strange. The English officers experienced it like that too—that thanks to soccer and Christmas, the feast of love, deadly enemies briefly came together as friends."

//////

The story of the "Christmas Truce" cuts to the unique power of sport to transcend language, politics, upbringing . . . anything. A British infantryman might try to kill a German gunner on most days, but they can still kick a soccer ball to each other. A goal, *golazo*, or *tor*—all are worth one point; it doesn't matter what you call it. In even those most extreme of times, sports can be the thing that unites us.

By this point in the book, you might be wondering how there can be a unified Religion of Sports when every team has a bitter enemy. Duke University and the University of North Carolina participate annually in something like a holy war. Maybe you're a Los Angeles

Lakers fan (I would recommend that you reconsider every one of your life's decisions), and you scoffed all throughout the first chapter of this book about my beloved Boston Celtics. The thing I've come to realize, though, is that it's not as if there is a Religion of the Red Sox and a Religion of the Yankees. We all believe in the same Religion of Sports, in the same teachings of sportsmanship, possibility, and wonder. The difference is that we are all just a part of different tribes.

Unlike many other areas of life, sports often show tribalism at its best and most healthy form. We settle our differences on the field. We shake hands after games. Rules apply, and winners get crowned democratically. Best man wins. Ball don't lie.

I have traveled all over the world reporting on the games we play, and one thing that always surprises me is how you can find members of your particular tribe anywhere. I've watched Celtics games with fellow diehards in Mumbai, Sydney, and Dubai. We might not speak the same language, but we share a common set of beliefs: we know #33 means Larry Bird, and that green is the holiest of all the colors.

Every now and then, though, sports can go one step further. It can achieve something like what the Germans and English experienced in the middle of war back in 1914. It can unite us, remind us of all the common ground that we share. When tribes come together like this, there's no stronger force in the entire world. There's so much good channeled together that for a moment, it seems as if anything is possible.

And I do mean anything. Thanks to the Religion of Sports, people from every type of tribe can be connected.

///////

There are few people in this world for whom I have more respect than Johnny "Joey" Jones. There are also few people with whom I have so little in common. We are each other's opposite in just about every way. Joey was born in a small town in northwest Georgia. I'm from Boston. After high school, Joey joined the U.S. Marine Corps. I went to Columbia University. Joey is a frequent contributor to Fox News. I watch MSNBC.

Joey loves NASCAR. For the longest time, I thought those races were a joke.

But NASCAR brought us together. We met in 2016 when I was filming a series for DirecTV also called *The Religion of Sports*. (A company, a series, and now a book? Maybe I need to start thinking of different names . . . but hey, if it works, it works, right?) The idea was that each episode would showcase a unique sports culture from around the globe. We had a full slate of travel planned to film the first season: Yosemite National Park for a mountain-climbing expedition; Calgary, Alberta, for the Stampede; Glasgow, Scotland, for the famous soccer derby between Celtic and Rangers. To start, my production team and I figured we'd get the episode we were *least* excited about out of the way first: an exploration of all things stock car racing.

As we researched potential characters for the episode, one of my colleagues, Adam, learned about Joey, a massive NASCAR fan. In the Marines, he defused bombs—real-life *Hurt Locker*. He served for eight years in both Iraq and Afghanistan. One day in 2010, his platoon reached a town in Afghanistan that was wired with improvised explosive devices, or IEDs. Joey and a partner went to work clearing out the city, defusing more than forty bombs, when Joey leaned against a wall to take a break. They had just a few more left, and once Joey had caught his breath, he took a step forward to get back to work when—*BOOM*.

He had stepped on an IED. The next thing he remembers, he woke up in a military hospital in Germany. Two days passed. Both his legs were gone, amputated above the knee. Remarkably, though, he was still alive.

Rehab was grueling, but Joey, as always, was not going to be anybody's burden. "As scared as I was, the people around me were even more scared," Joey told me. So every day, when he would get his lunch served at noon, he kicked everybody—nurses, doctors, visitors—out of his room. For twenty minutes, he cried and screamed. He cursed at God. He threw things at the wall. Somewhere in the middle of all that, he ate his lunch. Then, he would call everybody back in the room, smile, and tell them everything was okay, everything was going to be okay, why wouldn't it be okay? He told them he was going to walk again. Then, finally, as Joey explains it, "There was a day the tears didn't come. I wasn't mad anymore." That's the type of guy Joey is. You can see why I respect him so much.

While he rehabbed at Walter Reed National Military Medical Center in Washington, D.C., NASCAR star Kyle Busch visited him. Unlike most athletes, Busch visited a second time, then a third, then a fourth. The two became friends. By the final visit, conversation came easily. "You know what," Busch told Joey, "when you're out of here, you're coming to a race with me first thing."

Joey nodded. "And after I come to that race," he responded, "you're coming hunting with me."

So when Joey got out of the hospital, he visited Busch for a race, walking on two prosthetic legs. A little while later, Busch met Joey for a hunting trip. The two remained close. They road-tripped together in an oversized RV. Joey, a lifelong NASCAR fan, visits Busch on race day and sits near the mechanics, next to the pit stop.

We reached out to Joey to see if he'd be our guide through the world of NASCAR. He agreed, and I wasn't really sure what to expect. He'd show us where he grew up and introduce us to his family. Then we'd go to a dirt race, and finally, the Coca-Cola 600 on Memorial Day weekend at Charlotte Motor Speedway. As I boarded the plane, I remember thinking, *If I'm about to spend so much time with this guy, I just really hope nobody brings up politics.*

////////

Joey grew up at the southern end of the Appalachian Mountains, near Dawsonville, Georgia. Nearly a century ago, the area's quiet, mountain nights would be interrupted by the roar of engines. It's why Dawsonville calls itself the Birthplace of Stock Car Racing. During Prohibition, moonshiners needed to move product, and they needed to do so while staying as inconspicuous as possible *and* being able to outrun federal agents. So men drove their normal, everyday cars into the family barn and began to tinker. They removed floorboards, seats, anything that took up valuable weight and storage room. They added souped-up engines and extra suspension. These cars could fly, and throughout the night, that's exactly what they would do, boys as young as nine driving moonshine through the hills and around hairpin turns.

Soon the moonshiners started talking among themselves. One guy would say his car was the fastest in the county. "You've got to be kidding me," another would respond, and soon enough, local fairgrounds hosted stock car races. In 1937, twenty thousand fans came to watch a race, but during World War II, official races got suspended. When they finally returned on Labor Day in 1945, the police arrived at Lakewood Speedway in Atlanta to announce that five racers were ineligible to race—they'd been convicted of bootlegging. One of those racers was the legendary Roy Hall, an early racing star who had been arrested sixteen times. With that kind of record, Hall didn't even have a driver's license; it had been revoked. The crowd chanted Hall's name, thirty thousand of them strong, until finally the police agreed: the race could continue as planned, but the men in question would go to jail afterward. Hall won, and after he served his time, he typically entered races using a pseudonym.

I sat in the back seat holding a camera as Joey drove through those same Georgia roads that once roared with bootleggers. We were heading to Dawsonville's famous Pool Room—a restaurant and bar with a pool table in the middle. The drive was gorgeous, green trees everywhere, and it was easy to imagine how legendary drivers like Junior Johnson could turn off the road and lose the cops by taking a shortcut through the woods. "If you're a racing fan at all on the East Coast," Joey explained, "you've probably made your way to the Pool Room—especially if you're a Ford or Bill Elliott fan." Elliott was from Dawsonville, and as Joey told me, "Growing up, Bill Elliott was the biggest celebrity known to man." While racing in the 1980s, Elliott set track records for qualifying speed at Talladega (212.8 miles per hour, the fastest qualifying speed in NASCAR history) and Daytona (210.4 miles per hour), two tracks that even I know well. Elliott's son, Chase, races now, too, and whenever a member of the Elliott family wins a race, the Pool Room's owner, Gordon Pirkle, sounds a siren that can be heard all over town. In 2020, when Chase won the NASCAR Cup Series Championship, the siren rang for hours, and hundreds of people drove to celebrate.

We walked into the Pool Room, and the first thing I noticed was the floor. It was black and white, made to look like a checkered flag. The whole restaurant was a NASCAR shrine—newspaper clippings all over the walls, trophies behind the bar, an entire car's hood hanging from the ceiling. "I have not been in here in ten years," Joey told a waitress when she came to take our order.

"Well, nothing's changed in the twenty-eight years I've been here," she quipped back.

I sensed that this was a special place, somewhere worthy of the pilgrimage. I asked Joey for some more context. "The first time I came here, I was probably eight or nine years old, and I could have gone to Disneyland." That's how wide-eyed it made him.

The specifics of everything around me—the cars, the accents, the lore—was unfamiliar. But the sense of awe felt completely recognizable. I told Joey some of my stories from visiting Boston Garden, and he nodded knowingly. "Those guys in the race cars," Joey said. "They were heroes. They were superheroes." He was talking about Bill Elliott, but I could have said the same exact thing about Larry Bird.

I was starting to understand—and I hadn't even watched a race yet.

////////

After the Pool Room, we drove back through those hills to Joey's hometown of Dalton. Joey's whole family was there. It was dark, and we sat outside their home while the Joneses traded stories about racing. His grandfather wore a National Rifle Association cap, which added to me feeling like I was a fish out of water.

Turns out, Joey first started driving when he was twelve. His grandma put him in a Ford Thunderbird. He could barely see above the steering wheel, but then again, his grandma could barely see over the dash. She laughed that night years later when Joey told the story. She wore a racing shirt; their love of the sport transcended generations.

The Joneses aren't the only family where the same can be said. We interviewed Kyle Busch during our visit, and he told us that his first race ever was against his father. "If it's in your blood, it's in your blood," he explained. The most famous father-son racing duo in the sport is the Earnhardts. Yet what most people don't know about Dale Earnhardt Jr. is that his late father wasn't his only connection to racing. His grandfather was also a driver. On his mom's side, his grandfather was a body man, and his uncles were mechanics. Junior told us that as a kid, he'd visit with those mechanic uncles after races, and he'd run his hands along the cars while they told him what was wrong with them. Eventually, Junior could feel a car and, without having watched the race, tell them how well the car had run.

If it's in your blood, it's in your blood.

Joey raced as a kid. So many kids in this part of the country do.

Where I'm used to spending weekends driving all across Southern California to take my son to his next taekwondo tournament, fathers in this part of the country spend all week working on cars with their sons and daughters. On Saturdays, they load them onto trailers and take them to the dirt track. Kids who don't have their learner's permits win championships. "You need something that brings your community together," Joey said. "You need something that brings joy to your community. And there's a lot of places where that's church. There's a lot of places where that's religion. But for this part of the world, for where I grew up at, your church is on Saturday night."

It was a Friday. I asked Joey if he had any plans for the next night. I wanted to see one of these races for myself.

////////

The dirt track was on the outskirts of town, with grandstands like a midsize high school football stadium. It smelled like gasoline, and all over the infield, families worked on their cars. These automobiles weren't wrapped with Monster Energy and M&M advertisements like they are during NASCAR races. They looked like a scout's pinewood derby entry on steroids. They were spray-painted every color, windows busted out and replaced with netting. Some had busted fenders and others looked like it was their first race—although, judging by the care the mother was taking to check each tire, I wondered if it wasn't just a really good repair job.

I met a girl named Alexus Motes, whose blond hair was held back in braids and who spoke with the assured confidence of a teenager who knew she was speaking to somebody who *totally* didn't get it. She was one of the only girls at the track, and she got into racing thanks to her dad, David. "Alexus is usually the only girl in the race," David told me, "and I love it." Her car was white, with some green accents and a number 12 painted on the side. Above the windshield she had written BLACKJACK MOTOR SPORTS, a name that she and her dad made up to formalize their partnership. David wasn't like the overbearing parents whom you see so frequently at youth sports events across the country. Nobody at the track that night, at least among those I saw, was like that. These cars were labors of familial love, a way to link one generation to the next. And the act of racing? That was just a local rite of passage.

Joey spoke with Alexus and David about upcoming NASCAR races. Joey talked about his days on a dirt track and gave Alexus some tips. They were part of the same tribe.

Alexus and David left to finish preparing the car, and Joey and I grabbed a spot in the bleachers. A man who introduced himself as "The Faster Pastor" grabbed a microphone. "Even though I am a pastor in Charlotte, North Carolina," he said, "make no mistake about it, I was raised in the small town of Kannapolis, North Carolina, and I also raced in the Baja 1000. So, I understand what's happening here tonight: in my opinion, the greatest sport in the entire world and the greatest fans." Baja 1000? What on earth is that? Doesn't matter—it brought the crowd to its feet.

The race began, and a roar filled the air. I have to admit, it was exciting. I tried to pick out Alexus in the race, and at one point another driver tried to press her against the inside wall. I didn't know the strategy here, but I knew that wasn't good. She slowed down a bit, then zipped ahead. I let out a cheer.

The race lasted about twenty minutes. Win or lose, kids climbed out of their cars after the finish line and shared high fives or hugs with their parents. Then they checked under the hood for any damage and rolled the car back onto the trailer. The real repairs would be done at home, after schoolwork was finished. Then they'd bring it back next Saturday for another race. We waved goodbye to Alexus and David.

"I see you!" Joey said to her.

//////

NASCAR races are held on Sundays, right after church lets out. Many of the hundreds of thousands of fans who fill racetracks are still in their Sunday best; their pilgrimage to these races is part of their weekend ritual. Faith is tied to NASCAR drivers, too, and each week, immediately following the mandatory pre-race meeting that goes over track conditions and rules, a congregation of drivers, mechanics, and officials gather in a meeting room for a literal church service. Since 1988, an organization called Motor Racing Outreach facilitates chapel for anyone in the NASCAR community who wishes to join. Many of the biggest names in the sport attend the services, and when that ritual is finished, another one begins. Mechanics run to check the temperature of an engine. Racers study strategy. A different type of church is about to take place.

But first—one more prayer. In most American sports, the singing of the National Anthem symbolizes the beginning of competition. NASCAR features "The Star-Spangled Banner" as well, but uniquely, an invocation precedes the anthem. Television networks carry this prayer before every race; it's the only sport where TV networks show these complete pregame ceremonies before every event. It's been this way since the very beginning. In 1979, when CBS aired the first complete NASCAR race on national television, the Daytona 500, producers found the spectacle interesting and decided to air the whole thing. Nobody—except those who made the pilgrimage—had ever seen Daytona for themselves, after all. When executives surveyed viewers following the race, almost everyone who responded mentioned the invocation, and almost everybody liked it. So it's stayed ever since.

Often, they are rather routine prayers for health and peace. But there's one invocation that has become so infamous that even a NASCAR skeptic like me has heard about it. In case you missed it, this is what Pastor Joe Nelms said before a Nationwide Series race in Nashville, Tennessee, in 2012. I should remind you—this is 100 percent real. It's not out of *Talladega Nights*:

> Heavenly Father, we thank you tonight for all your blessings you sent and in all things we give thanks. So we want to thank you tonight for these mighty machines that you brought before us. Thank you for the Dodges and the Toyotas. Thank you for the Fords. Most of all we thank you for Roush and Yates partnering to give us the power that we see before us tonight. Thank you for GM performance technology and RO7 engines. Thank you for Sunoco racing fuel and Goodyear tires that bring performance and power to the track. Lord, I want to thank you for my smokin'-hot wife tonight, Lisa. And my two children, Eli and Emma, or as like to call 'em, the little E's. Lord, I pray you bless the drivers and use them tonight. May they put on a performance worthy of this great track. In Jesus's name, boogity boogity boogity, Amen.

Boogity boogity boogity! Have you ever heard a pastor speak like that? For years, I'd seen NASCAR on the TV and promptly switched the channel. What could be so exciting about four left turns? We traveled to the Coca-Cola 600, held annually on Memorial Day weekend at Charlotte Motor Speedway. We were on the infield, and all around us, families and old friends drank and grilled outside their mobile homes. They would travel from race to race, making pilgrimages to visit all the most sacred spots on the racing circuit. I wondered if they ever stopped by the Dawsonville Pool Room. We rode past the scene in a golf cart, which was sort of funny. Here I sat, in the middle of this sea of horsepower, riding the full throttle of a club car. We certainly weren't going fast—partially because of our ride and partially because

Joey seemed to know everyone. We started and stopped like a car trapped on the 405. He said hello to tailgaters outside their mobile homes, greeted security guards by name, and got fist bumps from pit bosses.

We settled near Kyle Busch's team to watch the race with Joey. The speed and care the pit crew showed in maintaining these machines stunned me. But it felt familiar, too. Although they were quicker and more technical, these professionals weren't all that different than what Alexus and David were doing at the dirt track. The pit crew rolled Busch and his car closer to the starting line. It was race time.

A pastor read a prayer, tens of thousands sang the anthem, and a man yelled, "Start your engines!" I'd been handed some earplugs, but I wasn't prepared for the wall of sound that hit me. This was a full-body type of loud. And right then, as the engines whirled to life and a surge of power filled the stadium—it all made sense. I put a hand to the ground and could feel it shaking. This was about connecting to something larger than yourself, something more powerful. Men and women trying to push their limits: the terminal velocity of a human being in free fall is about 122 miles an hour. That's our natural limit. But Joey's hero, Bill Elliott, worked these machines so deftly that he experienced what it feels like to rocket across asphalt going 212 miles per hour. He pushed the limits of man. That's what everyone was here to see, and all the preamble, the pilgrimages in the infield, the prayers—all of it was a celebration of this unique connection between man and machine. Of course Joey was drawn to this. "There are very few people who understand the welding of human and machine like an amputee. I've got two feet of machine under me as I roam this earth," Joey said. "But NASCAR drivers understand."

The cars raced off, and I whipped my neck to keep up. Joey and I made eye contact, and although we never said anything, he knew that I understood.

////////

Joey and I keep up. Often, it's on Twitter. I'll bemoan the state of the country from a Democrat's point of view, and he'll fashion a response. We've had a back-and-forth that lasted over a week. It's exactly the type of interaction I try to avoid online, but with Joey, there's always some civility. Most of that is because of the type of guy Joey is, but some of it is because of those days spent on racetracks. I wasn't a part of his tribe, but mine believes in a lot of the same things. We each have prophets, rituals, and lore. We understand each other in that primal way.

In February 2017, our paths crossed again. My Patriots were playing Joey's Atlanta Falcons in the Super Bowl at Houston. I texted Joey some friendly trash talk, which he returned, and then asked if I was going to the game. I was. He would be there, too. We made plans to meet up.

On game day, we met outside NRG Stadium. I introduced him to my family, and he poked fun at Krishu's Tom Brady jersey. We went to our seats, and at halftime, when the Falcons were ahead 28–3, I checked my phone. Of course, more trash talk from Joey. That's what fans do.

Then Brady took over the game. The lead shrank and shrank until the game was tied and headed to overtime. Brady led the Patriots down the field, and the Super Bowl was ours, an absolutely epic win. I immediately wanted to rub it in to Joey, but my barbs didn't seem to affect him. He was just as amazed as I was. "They deserve it," he said.

"I hated seeing the Falcons lose," Joey told me a few years later. "But man, look at that Tom Brady. Look at what he did. Look at Bill Belichick. Look at what he did. And so, all the hatred in the world for the team that beat the Falcons is still overshadowed by the respect I have for the people doing it."

There might be no more obvious example of tribes in modern life than sports teams. But overwhelmingly, these allegiances are forces for good. They are the community that fallen warriors can lean on to recover from a horrific injury. They are the traditions that can connect generations. They are the rituals that give the calendar a healthy rhythm.

When members of different tribes encounter each other, the shared beliefs in the Religion of Sports just might help them see past the fact that every other affiliation in their lives would keep them pulled far, far apart. Even when one tribe beats the other, the respect for the game, the inspiration for greatness, can transcend those labels. There's nothing left, at that point, but respect and understanding.

"That," Joey said, "is something we can learn from sports that we haven't used in our daily lives in a long time."

BILL ELLIOTT

NICKNAME: Awesome Bill from Dawsonville

DESCRIPTION: NASCAR driver, 1975–2012

FACT #1:

A fan favorite hailing from the small town of Dawsonville, Georgia, Bill Elliott ran more than 800 races in a storied thirty-seven-year career, which included a 1988 Winston Cup championship and two Daytona 500 victories. Elliott won NASCAR's most popular driver award a whopping 16 times.

FACT #2:

Elliott got his start driving a car owned by his father and with a pit crew operated by his brother. The Elliott family was a racing family through and through, and Bill's son Chase won the NASCAR Cup Series championship in 2020, making the Elliotts the third father-son duo to both win NASCAR championships, alongside Lee and Richard Petty and Ned and Dale Jarrett.

WISDOM:

"Daytona is a special place and the 500 is a special race. I haven't reached the point where I am content to watch this race on television. I would rather be part of the action. It's the big show." —*Bill Elliott*

CHAPTER 9

Afterlife

Making sense of it all . . .

If I have to face myself and say, "You're a failure," that's almost worse than death.

—Kobe Bryant

THE BUDDHA NEVER wrote anything down. The reason wasn't that he was illiterate; he was a prince, after all, and received an extensive education. Rather, the Buddha believed that his teachings could be best preserved orally. He traveled and taught, and as people listened and learned and watched, they would go and tell others what the Buddha had told them. Those people in turn told another and so on and so on. Soon the teachings passed from person to person, all across Asia, and then across the world.

This way of teaching the faith, from teacher to student, is central to many forms of Buddhism today. It's especially important for monks. If someone wants to become a Buddhist monk, he or she will seek out a guru, a spiritual guide, to teach them. The student will study the master, and then the student takes those beliefs and applies them to his or her own life, pushing the wisdom forward for the next generation. That way, when the next person comes along who wants to learn, there's now a new teacher to pass along the wisdom.

The result of this system is what is known as lineage. You can trace many of the great Buddhist teachers back millennia, a daisy chain of knowledge where each generation takes the lessons of their predecessors and pushes them forward, just a little bit, in his or her own way. If you go far enough up the chain, you can trace all wisdom back to the Buddha himself. "The qualification of a spiritual mentor," explains Naropa, a legendary figure in Tibetan Buddhism, "is that he possesses the lineage." Another scholar described lineage as "the unbroken rosary of Buddhas."

Buddhists believe that when they die, they can be reincarnated. Often, this means somebody returns to life as a more enlightened being. If someone spent their life studying and striving to be like one particular teacher, for example, then they can come back to earth just as wise as that guru. And then, in that next life, they can strive even higher—until they reach enlightenment.

There's something beautiful to me about this cycle of learning and teaching and striving to be better. There's something profound about the idea of received wisdom, that we can use our passion and curiosity to ensure that we can continue to live on after death—and that our beliefs and contributions will be passed on to the next generation.

It all reminds me of my friend Kobe Bryant.

///////

It is not fair to call Kobe Bryant one of the brightest athletes ever. He's one of the brightest *people* I've ever been lucky enough to know—curious, gracious, and creative. We were first introduced by a mutual friend, named Ellen, in 2013. Ellen knew I was a huge sports fan, and estimated Kobe and I were about the same age and might have common interests. What she didn't know was my intense loyalty to the

Celtics. When I brought it up to her, she shrugged it off. "I think you and Kobe will get along."

The Celtics and Lakers were just a few years removed from an epic, three-year run in which they met in the Finals every year. Those Celtics teams—with Kevin Garnett, Ray Allen, and Paul Pierce—could have competed against any team in NBA history, but Kobe seemed to be playing for something greater than just basketball. We won the first year, in 2008, but we could never beat him again. Every Celtics fan I knew would watch those games and mutter curses in Kobe's direction through the TV. We couldn't stop him. You almost had to respect the guy. Almost.

Now, here I was, face-to-face with this character who had ripped out my heart over and over and over again. I couldn't hold my tongue when we sat together at dinner. I knew it wasn't true, but I told him that his Lakers weren't *really* better than my Celtics. I told him he'd gotten lucky. Big man Kendrick Perkins's injury in Game 6 of the Finals was the real reason the Celtics came up short in their Game 7 loss to the Lakers. Metta World Peace's miraculous three-pointer that sank the Celtics in the final minutes was the luckiest shot in the history of the NBA. Classic trash talk. Somehow, I never learned the importance of picking on someone your own size.

But Kobe loved it. He talked right back, rubbing salt in my wounds. Sure, he was a Hall of Fame baller, but he was maybe the greatest smack talker of all time. Eventually, the hostilities subsided, and we started to bond over our shared love of hoops. We talked about our favorite players and teams and coaches. We talked about the up-and-coming talent in the league ("Do you think KD, Russ, and James Harden can win together in Oklahoma City?"). We also talked about storytelling. Kobe said he was interested in learning more about documentaries and eventually envisioned himself moving onto that field once his basketball days were over. He said he wanted to stay in touch and asked for my phone number—he might call me sometime in the future. Of course, I gave it to him, but I also assumed that I'd never hear from him again.

Just a few days later, an unknown number popped on my phone. It was Kobe. "Why don't you come down to Newport?" he said. "I want to show you something." Kobe lived in Newport Beach, about fifty miles from my home in Los Angeles, and when Kobe Bryant invites you to visit, you have to say yes. I got ready and was just about to leave when the phone rang again. It was Kobe's assistant this time.

"Hi, Gotham," she said. "Something's come up. We're going to have to reschedule. Kobe was wondering if you could join him for dinner."

"Sure," I said. "Any time. Just let me know which day works best for him."

"No," she said. "He was wanting to have dinner tonight."

"Oh. Okay. Uhhh . . . sure. I can move some stuff around."

"Meet him at the Chateau Marmont."

"Okay."

"Seven thirty."

"Got it."

"Oh, and Gotham?"

"Yes?"

"It's Kobe's birthday today." Then she hung up.

Kobe's birthday? What do you wear to Kobe's birthday dinner? What do you bring? Who else was going to be there to celebrate? I arrived at the Chateau Marmont early, half to make a good impression and half to make sure I wasn't getting punked.

I walked up to the maître d'. "I'm here with . . ." I looked around, leaned in, and lowered my voice. "Kobe Bryant."

She scanned the list of reservations, as if she didn't know the name, a move that always struck me as a little funny. Then she said, "Ah yes. He requested a table in the corner. He should be here soon. Follow me."

She led me through the restaurant. I looked around for big, long tables puzzled together for a group. But she kept walking past all of those and to a small two-topper in the corner. "Here you are, sir," she said.

The table was set for two.

Kobe arrived shortly after. We sat there, just the two of us, sharing a meal on his birthday. He quizzed me about storytelling, and I quizzed him about basketball. He was obsessed with GOATs in all fields, and so we talked about Elvis Presley, Michael Jackson, and Jay-Z. We had fun, and then some other friends joined us around dessert and we all shared more stories. Then I noticed Kobe looking at someone a few tables away. It seemed like he was getting nervous, jumpy almost, the way most of the diners had been acting earlier in the night when they first noticed *him*. I searched to see what or who had caught Kobe's eye when I saw her, sitting a few tables away: Katy Perry.

Kobe told me something about how much his daughters Natalia and Gigi loved the pop star. He said they would be so excited if they saw her there. "Go say hi," I told him. "Tell her that."

But Kobe refused. "I can't do that," he said. "It would be super-embarrassing. I don't want to bother her."

Here he was, the biggest star in the City of Stars, and he was utterly starstruck. In that moment, he wasn't Kobe Bryant. He was just a Girl Dad, excited to be in the presence of someone his daughters idolized.

We kept talking, about Katy Perry and other greats, and I wished him a happy birthday. I said good night and walked back through the restaurant, past rollicking tables of four and six and eight, each one of them celebrating something.

///////

Kobe called again the next day. "I still have something I want to show you. Can you get down to Newport?"

I met Kobe at his office. Books and papers were scattered about his desk, and in the middle of the mess was a framed photo of the whole Bryant family: Kobe, Vanessa, and their two girls. On the wall were four large, framed black-and-white portraits: J. K. Rowling, Walt Disney, Steve Jobs, and the composer John Williams. He waved me in. "Gotham," he said. "Come here. Look at this."

He reached toward a bookcase crammed with VHS tapes. Their spines had faded pieces of paper that long ago recorded which game each tape held. This was a catalog of all the great NBA matchups of the 1980s and '90s. I would come to find out later that when Kobe grew up in Italy, his uncle sent him these tapes so that his nephew could follow the NBA from overseas. Kobe had studied these videos, dissected every frame. Now he scanned the shelves for the right tape, hovered over one that said CELTICS-LAKERS, and pulled it out.

"Larry Bird," Kobe explained, "wasn't the strongest, he wasn't the biggest, and Lord knows he wasn't the fastest player in the league. But do you know what made him so great?" He held up the tape. "His pump fake. Watch this."

He put the game on, rewinding until he found a perfect example. Kobe didn't go backward and forward several times to find the clip; he knew the exact time stamp of the play he wanted to show me. He went directly to it. "People think it's his shoulders or his knees," Kobe continued. "It's not. It's his eyes. Watch his eyes."

I had watched Larry Bird since I was a kid. I had seen his greatest moments on ESPN Classic so many times that I had memorized his every step on the court. But I had never noticed his eyes. Whenever Bird went to pump-fake, his eyes focused on the hoop just as intently as when he was actually shooting the ball. That's what a defender looks for, Kobe explained. A defender doesn't care about how fancy a shooter's footwork is if they're trying to determine whether a shot attempt is real. They look at the eyes, and when they looked at Bird's eyes, there was simply no difference between shots and fakes.

Kobe grabbed another tape, a different Celtics-Lakers game from that era, and put it on. Kobe knew everything about every player on the court—not just the Larry Birds and Magic Johnsons, but the Scott Wedmans and Mike McGees, too. We watched the two rivals duke it out for hours, continuing the previous night's trash talk. Then he put on an Oscar Robertson game. "See how he uses his body?" Kobe said. "Now look at this." Kobe pulled up a game of his own, and as he drove toward the basket, bodying players for better positioning, it was like watching a frame-by-frame re-creation of Robertson.

I asked him about Michael Jordan. It was an open secret that Kobe modeled his game after MJ, and on YouTube, you can find spliced-together pictures of MJ going up for a reverse layup, tongue stuck out, followed by Kobe doing exactly the same thing while making exactly the same face. It wasn't the style that Kobe copied, he told me. At least not intentionally. He put in another tape. Jordan in the Finals. "Look at his footwork." Sure enough, when we went back to look at Kobe's own highlights, it was like watching a mirror.

Kobe had watched these tapes over and over and over again. In many ways the story of Kobe Bryant was less about him and more about his fascination with others. Pretty soon, I realized that Kobe hadn't invited me there to get the last laugh in our trash talking, but because he wanted to keep grilling me about storytelling. "I want to get into content," he told me. "I want to tell my stories someday." He had this idea for a documentary where he would talk about all the disparate figures from whom he drew inspiration. He had studied the four people whose portraits decorated his wall, he said. He'd even picked up the phone and called composer John Williams once to ask him how he created the theme for *Star Wars*. There were others, too:

he had called Anna Wintour, Arianna Huffington, Hilary Swank, and countless others. Kobe had learned about what they were doing, thought it was interesting, and wanted to know more. And because he was Kobe, he wouldn't stop until his curiosity felt appeased. Now he wanted to find a way to channel all these different thoughts together. His basketball career was in its final act. He wanted a way to reflect on everything, to celebrate the journey he had taken.

I told him that the idea was fascinating—and it was, truly—but that he couldn't make a documentary just about inspiration. There had to be a story to tell. There had to be some structure. What was the narrative?

Kobe seemed to understand. He said we'd keep in touch. We texted occasionally, most of the time more trash talk about the Celtics and the Lakers. And then one day in April 2013, while I was tracking down Sachin with Victor, my phone started to explode with texts in the way that happens only when earth-shattering news occurs. Kobe had ruptured his Achilles. A friend sent a video of Kobe's postgame press conference. Cameras and a phalanx of microphones cornered Kobe against the wall. He was still in his jersey. He could barely look anyone in the eyes; clearly, he was in pain both mentally and physically. "I can't walk," he said. Reporters kept grilling him. "Players at this stage of their career, they pop an Achilles, and pundits say they'll never come back the same," Kobe said. "I can hear it already. It's pissing me off."

Another reporter asked, "So this isn't the end?"

Kobe gave him a side-eye. "Really?" he said. Give up? Had this guy never heard of Kobe Bryant?

I texted Kobe right away. "This is the spine of the project. This is the story." One final comeback.

"Good idea," Kobe replied almost instantly. "Hit me up when you get back in town."

///////

When I made it back stateside, my camera and I started to follow Kobe everywhere. To physical therapy. To midnight gym sessions. To the doctor's visit where Kobe was told that his injuries would affect him for the rest of his life. "This injury," he told me, "was Mount Everest for me."

It had all the beats of a classic narrative—the comeback story!—and so I set off with my team to turn it into a documentary. We wrote a treatment, shopped it around, and sold it to Showtime. For eight months, when I wasn't with Kobe, I interviewed dozens of people who played a major role in his life: Phil Jackson; Shaquille O'Neal; Kobe's wife, Vanessa; his sister Sharia; and many, many others. At the end of every interview, I asked everyone for three words that came to mind when describing Kobe. Most of the responses were about what you would expect: "competitor," "intense," "passionate," "driven," "relentless." Vanessa had ones that gave a glimmer of a side that only she probably knew: "goofy," "tender," but, of course, even she described him as "competitive," and told great stories of how the two of them competed over, well, everything.

Then there was his former teammate Steve Nash. I asked him the question at the end of the interview. He paused for a moment. Then he looked straight at the camera and said with a smile, "Mother . . . Fucking . . . Asshole."

When I yelled cut, Steve turned to me: "Tell him I said that."

So, I did. Kobe laughed hard. "I love that! It's true! It's true!" Then I asked him the same question. *How would you describe yourself in three words?* He shrugged.

"I'm just me," he said.

///////

We cut all the footage together and created a rough cut. It followed a pretty typical documentary format, but I felt excited with the result. There was so much in there that hadn't been discussed before, so many great nuggets from some of the most iconic basketball players in history. I was proud of what we'd made, and so I drove down to Newport Beach and showed Kobe.

It's always odd, watching someone else watch something you created about them. You study their face to see whether they like it, or if a certain part makes them mad. There was a section in the film that dealt with a 2003 criminal probe in Colorado, where Kobe had been accused of sexual misconduct. The case didn't proceed when the alleged victim refused to testify. A civil claim was later settled out of court. I was especially tense observing Kobe watch that scene, just waiting for him to say something. But the tape kept rolling, and the room remained quiet. No complaints from Kobe.

While studying Kobe that day, it struck me that it must be strange to be the subject of a project like this while you're still playing, to hear your peers discuss what an impact you have had on their careers and lives. But if those thoughts ran through his head that day, he didn't show any signs of introspection. I couldn't tell what he was thinking; all he gave me was a poker face. "Let me think on it," he said when the credits rolled.

I drove home, and when my wife asked me what Kobe thought, I could only answer, "I have no idea."

I didn't hear anything for several days. Then, a call: "We should do more interviews—you and me. A lot of these people you have talking don't know anything about me."

I told him, "Of course." Sure. The hardest part of working with elite athletes is wrestling them down for their time, and now one of the greatest ever is saying he wants to get more footage? It was a dream come true.

We wrangled together a crew and filmed the interview as quickly as possible. Kobe was more open than he had been for any of our other on-camera discussions. I thought about how we could work the new material into what already existed and was getting excited. The new footage did nothing but make the film better.

Then, another call: "You know what? I want to do more. When can we film again?"

Then another: "How soon can you get down here? I have some things I want to add."

Yet another: "Actually, there doesn't need to be anyone else talking but me. Get rid of everyone else."

Get rid of everyone else? At first I argued with the creative merits. Then I tried to be practical. Having the documentary *only* feature Kobe would mean that we'd have to totally start over. We would need more footage, more edit work, and more time. Most of all, as I explained to Kobe, "We're already over budget."

"We'll get more," Kobe said.

"That's not how it works."

"Set up a meeting with Showtime."

"I can try, but you'd have to be there."

"That's fine."

I set up a meeting with David Nevins, then the head of the network. Kobe drove up in a Cadillac Escalade and walked through the office building still wearing his sunglasses. He looked, and acted, like a rock star. Kobe marched right into Nevins's office, and even this Hollywood power broker got a little starstruck. Kobe told him about his vision, about all the things he was going to talk about, about how great the film was going to be. Nevins was sold. "How much do you need?" he said.

Kobe leaned back. "I don't talk about money," he said and looked toward me.

It was up to me to figure out the details, and the negotiation was the easiest I've ever been a part of. We got everything we asked for.

We started to shoot these long, emotional interviews. Kobe, centered in the frame, talked about everything, including more details about the events in Colorado and how it impacted him and his marriage. They were some of the most intense shoots I've ever been a part of. When it came time to edit, to actually put the pieces together into a story, Kobe demanded that we set up postproduction near his home in Newport Beach. We built an entire edit bay there. Among the crew, it came to be known as "Kobe Jail."

There is no version of Kobe Bryant that is passive. As we worked on the film, he asked questions about everything. At first, it was endearing. "What does a DP do?" "How do you work with a composer?" "What's the best software for editing a project like this?" Soon, though, it came to be overbearing. Kobe would literally stand over the shoulder of the editor as we worked on a particular sequence, demanding, "I want that montage to be three seconds shorter." He stayed in that edit room until three in the morning, which means we all stayed there that long, too. One of my partners on the project rented a house in Newport Beach so that he could be available 24/7. I commuted.

During that time, I constantly reminded myself and the team about who we were working with. To figure out the best way to handle Kobe the cocreator, I thought I could draw some inspiration from his former teammates. He had unimaginable success with Shaq, and the two of them fought all the time. Whenever a rookie or free agent joined the Lakers, Kobe famously pushed them harder and harder, trying to force them to prove themselves. The best-known example of someone who couldn't cut it in Kobe's mind was Smush Parker, a point guard who hung on with the Lakers for two seasons and whom Kobe very nearly bullied out of the NBA. "You can either be Shaq, or you can be Smush Parker," I would tell my crew.

As we put together the new version of the film, Kobe had a problem with everything—every shot, every music cue, every sequence. Everything had to be perfect, and in his mind, nothing ever was. However, I wasn't ever sure if he *actually* had so many issues or if he just wanted to hear us defend every decision. One time I snapped at him: "When I need help with my jump shot, I'll come to you. But don't tell me how to tell a story." He jabbed back, but I could tell he appreciated the jawing. This was the environment in which Kobe thrived. He always had to prove something.

Once, early in the process when we were discussing what tone the project should take, one of our collaborators referenced Darren Aronofsky's *Black Swan*. The idea was that we could emulate the film's opening, particularly the way that it created a specific foreboding sense of intensity. The more we discussed it, the more everyone in the room seemed to get excited about the idea. We talked about why and how it would work. But Kobe, uncharacteristically, sat quiet. I asked him what he thought, and he just shook his head. He hadn't seen *Black Swan*, he said. I offered to pull up a link to the opening scene, and we could all watch it together.

"No," he said. "I'll watch it tonight, and we can talk about it tomorrow." I agreed to the plan.

The following morning, I met him at the edit bay. It couldn't have been more than twelve hours since we were last in the same room together with the group.

"So, what did you think?" I asked.

He nodded. He liked the idea. But he also wanted to talk about a specific scene in Aronofsky's *Requiem for a Dream*. And he really liked the edge and intensity of *Pi*. Truly, he continued, *The Wrestler* may be the best analogue. And while he knew *The Fountain* hadn't been entirely well received by critics, he liked the surrealism of it and thought there were probably elements of that film that we should think about.

I stared at him speechless. I was doing the math in my head. Those films totaled up to more than eight and a half hours of run time. It was quite possible that he had done nothing else since I last saw him but watch Aronofsky movies. And because it was Kobe, there was a better than zero chance that he had not even eaten a meal or gone to the bathroom, let alone slept.

I asked him why he felt the need to watch all those films.

He shrugged. "How could I have an opinion on *Black Swan* and why Aronofsky did what he did if I didn't understand his mind?"

That was Kobe. It was not about understanding the film. It was about understanding the mind of the artist who created it.

///////

When I think of Kobe now, I think of him less as an athlete and more as an artist whose medium just happened to be basketball. And basketball, I would come to learn, dominated Kobe's life. "The thing that was always the most constant was the game," Kobe told me once. "That was my refuge. That was the place where I could go to and have complete familiarity, no matter where I was."

The more we worked on the film, the more it became clear that what we were creating was less a traditional documentary and more a cinematic memoir. Our interviews became confessional for him, and it struck me that I had caught him at a particularly pensive part of his life. For the first time, he was confronted with the idea that he couldn't play forever. He had to consider what he would do after the NBA. What would he do with all the gifts that his talent had provided for him? Would he lead a rewarding life or constantly be tortured by the myth of the Black Mamba? Could he use the sport to set up his next chapter? What would he do during his basketball afterlife?

Many athletes struggle with these questions at the end of their careers, but for Kobe, the pain of losing basketball was particularly acute. It was like moving away from home for the first time.

We spent hours talking about Kobe's relationship with the sport. His father, Joe "Jellybean" Bryant, played in the NBA and extended his career by competing in Europe. When Kobe was six, the whole Bryant family moved to Italy, and there Kobe was, the Black American who barely spoke the language. He had trouble making friends. What he loved, though, was being with his father. Kobe slept in his clothes so that he could always be ready whenever his dad left for the gym. During his father's games, Kobe sat under the basket with one of those big, circular mops. Always obsessive, Kobe sprang up and started cleaning the court if even a single drop of sweat fell onto it.

His father never stayed on one team for more than a handful of years, and so every time the Bryants moved, Kobe had to start over with a new school, new friends, and another round of being the outsider. He found solace in two ways: by watching basketball and by playing it. Italian television didn't broadcast the NBA back then, so he turned to those tapes that his uncle sent him, the same ones he had shown me in his office. "It was a place I could go to and not be alone," Kobe said of watching those recorded games.

He would also play. He'd study the moves of the greats and try to emulate them. "Whether I got along with the kids or didn't get along with the kids, it didn't matter because I always had my ball," Kobe said. "I could always pick up my basketball, I could always hop on my bike, I could always go to the park, and I could always shoot. And that gave me a great source of comfort."

The Mamba Mentality, I learned, was born out of loneliness.

When he got back to the States, Kobe was still the outsider. He was the kid in his Philadelphia middle school who dressed funny and spoke Italian. He remembered sitting in the cafeteria eating lunch alone, Kobe Bryant starring in a John Hughes movie. It was confusing for a teenager. It was frustrating. Still, he had basketball. He always had basketball.

"I had all this resentment and anger inside of me that I hadn't really let out," Kobe said. "It was just never viewed as 'I'm going to control this thing.' It was more like, 'You know what? I'm just going to delay the eruption. I'm just going to push it to the side, and then use it for my benefit for what it is that I love doing, which is playing the game.'"

He continued, "And once I discovered that, everything about the game changed—because now, I understood that I could really lose myself through the game. No matter what affected me, no matter what happened in life, I could always step on the basketball court and let my game speak to that. I could step onto that court and just absolutely erupt. And that feeling of playing with that rage was new to me. But I fucking loved it."

The project had started as a discussion of Kobe's muses, and the more we talked, the more it became clear that although he drew inspiration from everywhere (he said once during rehab that if Beethoven wrote his Ninth Symphony while deaf, then he could play basketball with one and a half legs), his real muse was basketball. Kobe made a promise to himself in high school. He swore that he would become the greatest basketball player of all time. It became his mantra, his purpose for living. Everything he did was funneled through that filter: *Does this help me become the greatest of all time?* He had few friends. He practiced instead of partying. Pretty soon he announced that he would skip college to join the NBA. He wanted to learn from the best, and the NBA was his only opportunity to do that. Suddenly he was seventeen years old and standing next to Lakers legend and executive Jerry West, holding a purple and gold #8 jersey at an introductory press conference.

In this moment, again, Kobe became an outsider. His teammates would go out to clubs on the road, and Kobe had to sit in the hotel and order room service; he wasn't twenty-one yet and even the best fake ID couldn't hide the fact that bouncers knew exactly who he was. He told me a heartbreaking story once that several times when he was a rookie, he got in his car (he had only had his driver's license for one year) and drove around UCLA's campus. He went during the day and watched kids going to class. He went at night and saw groups coming and going from frat houses. "I just wanted to feel that," he said. He doubted himself, thinking, "Did I make the wrong choice? Did I fuck up?" Then he would drive away, leave that alternate reality, go to the gym, and keep shooting.

"The game," he explained, "became everything to me."

He continued, "I knew that I wasn't going to be stopped, because at the age of eighteen, this was my life. So you can't possibly become better than me, because you're not spending the time on it that I do. Even if you *want* to spend the time on it, you *can't* because you have other things. You have other responsibilities that are taking you away from it." Kobe, on the other hand, only had basketball. "So, I already won," he said.

Now here he was, sitting in a dark edit bay all these years later. Injured. Near the end of his career. What would be next? Kobe lived this mythic life, and the one constant through it all had been basketball. What would he do without it? I came to see that coming back from the Achilles injury was as much about getting healthy and proving everyone wrong as it was about trying to hold on to this relationship for a little bit longer, to get everything possible out of the sport. But even Kobe saw the end coming. "At what point do you feel like you're holding on, holding on, holding on to something that's not there?" he said. "At what point does your determination and your drive become something unreasonable?" He paused for a moment. "Or something that's just not possible?"

///////

Our documentary came out, and Kobe retired a season later. Of course, in his final game, he scored 60 points and hit a game winner. He took more than 50 percent of the team's shots that night, the first player in six decades to attempt more than 50 shots in the final game of his career. I thought, watching that performance, about a night in Newport Beach when we took a break from editing to watch an NBA game. In the first half, one player shot nine times and missed every single time. When the game ended, I checked the box score and saw that he never attempted a shot the entire second half.

Kobe shook his head in disbelief. "Bro," he said. "I'd go 0 for 49 before I stopped shooting the ball." I laughed, and he continued. "The only way you really lose in this game is when you quit and beat yourself."

I've thought about that moment a lot over the years.

Some athletes build business empires after retirement, but Kobe wanted to advance his storytelling ambitions. For some time, we talked about other projects that we could partner on. He didn't need me, of course. I was convinced that he was going to be wildly success-ful in media—maybe even more so than he was on the basketball court. After all, he won an Oscar just two years into his new life.

We spoke occasionally, and during those first few years, Kobe ad-mitted that he was struggling to figure out how to live without the rhythm and routine of the sport. He wasn't sure how to interact with the Lakers now that it wasn't his team. He wasn't sure how to fill his days.

Kobe found his answer, or part of it, through his teenage daughter, Gianna, known as Gigi. Gigi blossomed into a fierce basketball player herself, and Kobe helped coach her team. By seeing her fall in love with the game, Kobe found a new connection to the sport. Her passion fueled his. He felt more comfortable in his skin again. He showed up at Laker games—Gigi by his side.

He seemed to have it figured out.

It's why what happened next was even more heartbreaking.

//////

On Sundays, I play pickup basketball, from 9 a.m. to 11 a.m. That Sunday, January 26, 2020, was no different. I played the game, and checked my phone. A friend of mine who is a die-hard Lakers fan texted. "Kobe. Omg."

I assumed it was something basketball related. It must be good news. Was he coming back? Coaching?

I got into the car, turned on the radio, and started to hear unverified reports of a helicopter crash. Kobe on board. Rumors. I didn't believe it. It couldn't be true. Kobe Bryant couldn't die. I didn't realize it until that moment, but I always sort of assumed that he would live forever.

Kobe approached basketball the same way that a monk studies religion. He sought out gurus. He learned the lineage. He took what those who came before him had taught him, and he made it his own. At times, he seemed connected to something larger than himself— using his body like Oscar, pump-faking like Larry, moving his feet like Mike. Kobe lived for basketball, and he lived his entire life through the prism of sport. In return, basketball and its gods helped give him purpose.

"The goal," Kobe once told me, "was to sit at the same lunch table as my muses: Michael. Magic. I wanted to be able to sit down at the same table as them and belong there."

Kobe wanted to learn from his teachers—and then he wanted to become just like them.

In the days following his death, Kobe's presence became inescapable in Los Angeles. Every street corner, it seemed, had been decorated with a mural of Kobe and Gigi. Everyone wore #24 jerseys. We talked earlier in this book about how there are different tribes within the larger Religion of Sports. There's certainly a distinct tribe of Lakers fans, but within that, there's a Cult of Kobe, a group that grew up with him, that saw him live his life on a basketball court, that believed wholeheartedly in the power of the Mamba Mentality. The next time I played pickup, I wore a pair of sneakers that he had given to me, and even after I missed some jump shots, I thought about what he had told me about going 0-49. I shot again.

Two days after his funeral, I saw Kobe again. The Lakers were playing the Houston Rockets. LeBron James ran down the court at the Staples Center on a breakaway. He got the ball, dribbled once by the free-throw line, then exploded into the air, reversed the ball with two hands, then slammed it into the basket over the back of his head. In 2001, against the Toronto Raptors, Kobe made exactly the same move. It was on the same court, in the same spot, when he had dribbled by the free-throw line, exploded into the air, reversed the ball with two hands, and slammed it into the basket over the back of his head.

"Ever see the movie *The 6th Man*?" LeBron said after the game. "Kobe came down, put himself in my body, and gave me that dunk on that break."

In some sense, the way that Kobe lived his life through basketball truly does ensure that he'll live on forever. His jersey will always hang above the court, next to his banners. His records will always stand. He is one more link on the daisy chain. I can't help but think of Kobe, as a kid, getting boxes full of VHS tapes in Italy. He'd sit there and study them, and when he got to the park, he'd imitate them, and when he made it to the league, he used those skills to become one of the greatest of all time. As long as we play, basketball lovers from LeBron James to an American kid in Italy will watch him, study him, try to become him. They'll find solace through that quest, and they'll work hard.

One day, they'll get on the court, slam in a dunk, and find that they can't explain what just happened. They won't know how they did it. It was like they were channeling something more powerful than themselves. In that moment, they'll have felt the Religion of Sports. They'll have felt Kobe in the game. They'll have felt Kobe in basketball.

That's where he's always been.

KOBE BRYANT

NICKNAME: The Black Mamba

DESCRIPTION: Guard, Los Angeles Lakers, 1996–2016

FACT #1:

One of sports' fiercest competitors, Bryant defined his era of the NBA, winning 5 championships and making 18 All-Star teams.

FACT #2:

Scored 81 points in a 2006 game against the Toronto Raptors and put up 60 points in his final game before retirement, including a game winner with thirty seconds left.

WISDOM:

"It's the one thing you can control: You are responsible for how people remember you—or don't. So don't take it lightly." —*Kobe Bryant*

Afterword

IN THE FALL of 2021, my son Krishu got in trouble at his prestigious, progressive private school in West Los Angeles. We had meetings with his principal and a disciplinary board of teachers and soon learned that he would be suspended. Along with some other friends, he'd participated in something silly online. It was the exact sort of thing from which fourteen-year-old kids can learn valuable life lessons, but the school decided to crack down and punished the kids severely. My wife and I—along with all the other parents—felt betrayed by the school and profoundly disappointed by the way they handled the situation, but we also felt there was little we could do, especially if we wanted our kids to be able to return once their suspensions were over. It was the most helpless feeling I'd ever had as a parent.

The suspension came during the NFL season, when I was taking near-weekly flights across the country to film with Tom Brady as he tried to lead the Tampa Bay Buccaneers to back-to-back Super Bowls. When the cameras turn off on trips like these, Tom and I often catch up about our lives and our families, and so one week, when Tom asked me how Krishu was doing, I told him about the trouble and the suspension. He could tell I was upset by the whole thing and sad for my son.

"Are you coming back to Tampa soon?" he asked.

"Yeah," I said. "I'll be back in two weeks for the next home game against the Bills."

"Bring Krishu," Tom said. "I want to talk to him. We can drive home from the game together. It'll be fun."

So, in mid-December, I brought Krishu with me to Tampa, only telling him that it would be fun to watch the game in person. Tom had left us tickets to watch in his family's box. The Bucs took a commanding lead at halftime, 24–3, but nothing went right in the second half. Tom's sisters paced around the box. *Don't talk to me. Don't look at me.* Tampa surrendered 17 points in the fourth quarter alone, and before we knew it, the game was heading to overtime.

If the Bucs don't win this, I'm not sure I want to be sitting in a car with Tom Brady afterward, I thought, looking over at Krishu.

But I should have known better than to doubt #12. In overtime, the Bucs had the ball on third down near the 50-yard line. Tom hit receiver Breshad Perriman on an out route with nothing but open field in front of him. Perriman ran 58 yards to score the game-winning touchdown and make it Tom's 700th career touchdown pass. Tom's parents, sisters, and kids all leapt up cheering and hugging.

"Come here," I told Krishu, and I steered him to an elevator that took us straight to the area outside the locker room. We waited there for a little while as fired-up Bucs players streamed outside to hug their families. Gronk came out and, when he saw me, walked up to say hello. We'd worked together as part of our various projects through the years.

"Gronk," I asked, "have you ever met my son, Krishu? He's a big fan."

Gronk's eyes got big, his eyebrows raised. "*Yoooooo!*" he said, pointing at Krishu. "How old are you?"

Krishu answered, "Fourteen."

"Holy shit," Gronk exclaimed. "You're a big kid. Don't play football," he laughed.

He held his hand out for a high five, which Krishu returned, and then Gronk was gone like a hurricane continuing on its path. Krishu turned to me, not actually forming words but saying everything he needed to with his eyes, opened as wide as I'd ever seen them. Did that just happen? Did Gronk just bro out with him? Krishu was in disbelief.

A few seconds later, Tom came out.

We congratulated him on the game, made a little small talk. Then Tom turned to Krishu. "Jump in the car with me. I want to talk to you."

Krishu stared at me, even more wide eyed than before. I nodded, and we piled into Tom's Range Rover. I let Krishu ride shotgun next to Tom in the front. I sat in the back.

Tom weaved past stadium security, guards lifting cones for the quarterback to get through, and he slowed to thank each of them. *"I appreciate you, big guy!"* The humid Florida night caused the windshield to fog. Once we got on the road, Tom turned to Krishu. "Your dad told me about some of the BS you're going through at school," he said. Tom looked at me through the rearview mirror. "Can I swear with him?"

I nodded. Krishu had definitely heard far worse, especially over the last few days.

"Sounds like some fucking bullshit," Tom emphasized.

"Oh my God . . ." Krishu nodded, unsure how else to reply. I too sat in silence, determined to allow the moment to play out on its own.

Tom shook his head. "Listen, man, nobody expects you to be perfect, you know what I mean?" he said. "You're fourteen years old. You're *supposed* to make mistakes. That's what kids do."

Tom looked at Krishu for a response. But Krishu just sat silent, staring at the man who'd just led his team to an epic win and made NFL history in the process. Just in case Krishu forgot about the night's heroics, Tom's right shoulder was bulging with an ice wrap. Slowly, carefully, Krishu nodded. But he didn't say a word.

Tom admitted that when he was Krishu's age, he made mistakes, too. He confessed that he *still* makes mistakes, all the time. Messing up is unavoidable in life, he explained. "You just gotta use it as a learning experience," Tom said. "That's the most important part of going through stuff like that. You say you're sorry—and your dad told me you *did* say you're sorry—and then you learn and move on."

He continued, "My point is: Use this time to do the best you can with your family, and then learn from it. Learn a good lesson. Don't let one mistake define you. Don't let what happened make you feel any different about yourself. And Krishu?"

The car rolled to a stop at a red light, and Tom turned to look at my son. "I've known you since you were just a little shit. I know you're a really great kid. Don't forget that."

The greatest quarterback of all time was giving Krishu a Vince Lombardi–style pep talk during one of his lowest moments.

"By the way, I know what it's like to get screwed over by an institution, too. Effing Deflategate." We all laughed at that, and Tom pulled ahead driving into the night.

Since that drive, I've thought about the arc of my relationship with Tom. From the first time he and I talked about the Religion of Sports in Los Angeles, where I laid out my thesis that sports had all the elements of religion—from their various rituals, mythologies, sacred cathedrals, pilgrimages, and more. Over the years, Tom and I talked endlessly about this (along with Michael Strahan and our team at the company), forming a veritable scripture that became our ethos. The Religion of Sports was far greater than the sum of all those ingredients. It was something bigger than any of us. It was a code of ethics. A shared set of values. A way to live. It bonded players and fans, the greatest quarterback in history with a shamed fourteen-year-old kid. Looking at Krishu in the car that evening, I could see a burden had been lifted from his shoulders. He was at ease for the first time in weeks. I too felt a relief wash over me.

Afterword

When Krishu and I returned home to LA from Tampa, I shared the details of our car ride with my wife, Candice. And we decided it was time not just to talk about our values, but to live them, too. A few weeks later, once Krishu and his classmates were permitted to return to classes, we let our son's prestigious, so-called progressive private school know that we would not be returning the following year.

/////////

You probably already do many of the things necessary to be a true believer. That's why sports have so much power already, and it's one of the best parts of our faith: it's easy, and fun, to put into practice. Now that you've read about *why* sports are like a religion, I wanted to put together this guide to help you begin to understand *how* you exist as a member of this faith. In the pages that follow, you'll find some questions to ponder, and some activities to follow—all based on what you have read—in the hope that I can help give you a deeper, more visceral understanding of our shared love.

Sports is religion.

Believe.

The Playbook

Chapter 1: Baptism

What we learned: Sports—and the Boston Celtics in particular—were like a language to me as a kid. They helped me feel like I belonged in America even if my immigrant parents didn't understand why I was so obsessed with a big white guy named Larry Bird. I took my first pilgrimage to the old Boston Garden and witnessed a miracle when Michael Jordan scored 63 points in the playoffs. As Bird said after the game, "That wasn't a basketball player. It's just God disguised as Michael Jordan."

Are you a huge Lakers fan who fumed through my whole recounting of Celtics lore? Here's how you—and everybody else—can advance your faith.

It's good to start any spiritual journey with a little reflection. Take some time to meditate on the following questions:

- What team made you fall in love with sports? Do you remember your first game? Who was playing? Who was the athlete who seemed more like a god than a

human? What food did you eat as communion at the
stadium? What about it kept you wanting to come
back for more?

- Who was your guide into the world of the Religion
  of Sports? In this chapter, I called our family friend
  Alan my "shaman." Who played that role for you?
  Your mom? Your dad? A grandparent? Sibling? A
  high school coach? Every religion has evangelists—
  those who help new believers maneuver and navigate
  their newly discovered faith. Paying homage to them
  in the Religion of Sports is vital.

- In this chapter, we came upon the revelation that
  a religion is ultimately rooted in its community
  of believers, the shared experience of those that
  subscribe to a faith. It's with our flock that we
  dream up new scenarios for our favorite players
  and teams, imagining trades that will turn a season
  around, epic comebacks, and that sweet moment
  where we finally claim a championship. Pick one
  of your favorite teams. What would it feel like to
  win the championship next year? Where would you
  be watching? Who would you be standing next to,
  cheering alongside? These are your people.

Much like traditional religion, sports are inherited by each generation. A child is an Atlanta Braves fan because her mother grew
up adoring Dale Murphy and her grandparents remember when the
team moved from Milwaukee with Hammerin' Hank Aaron. I became
a Celtics fan because of the kindness of our family friend Alan, who
offered tickets and once-in-a-lifetime opportunities. So, I ask you . . .

How can you pass the Religion of Sports on to somebody else?

Let's proselytize for a little bit. Find one person whom you can invite to watch the next game with you. Sports are a community transaction, the shared experience around the thrill of victory and the agony of defeat. Spread the faith by extending the flock. The Religion of Sports can be such a powerful tool for good.

Chapter 2: True Believers

What we learned: There's a difference between saying you believe in a faith and living that faith as a true believer. Everybody knows that Tom Brady is the greatest quarterback of all time, but few people understand just how much he has dedicated to football. "In the end," Tom explained to me during the 2017 season, "my life revolves around football. It always has been, it always will be, as long as I'm playing. I've given my body, my everything—every bit of energy for eighteen years to it. So, if you're going to compete against me, you'd better be willing to give up your life. Because I've given up mine."

That's some serious commitment! Now, I'm not advocating for you to *totally* (or literally) give up your life for sports (although I suspect many of us have loved ones who would counter that we already spend *way* too much time obsessed with our favorite teams). But one of the great things about sports is that games, seasons, and entire careers can act as stages to reveal so many fundamental truths of humanity. As I've witnessed Tom's career as a fan, creative collaborator, and then as a friend, I've drawn some meaningful life lessons.

The Tao of Tom

- Control your inputs and your outputs. What you get in is what you'll get out—so be careful about bringing unwanted distractions, negative energy, and unfulfilling work into your life.

- Your greatest failures are also your greatest opportunities to grow. Watch the film. Study the losses. There's always something that you can learn from your losses.
- Even if everybody tells you that you cannot possibly get any better, defy the naysayers. You can always get a little bit closer to perfection. Even if you're only improving by 1 percent, work to improve yourself by that fraction, and one day, you might end up making that perfect pass. That's being an *Edger*.

See? There are a few little doses of the GOAT that you can carry into your everyday life—and you don't even have to eat pound after pound of kale.

Chapter 3: Myths

What we learned: Kathrine Switzer didn't set out to become a folk hero when she stepped up to the starting line in Boston in 1967. She just wanted to run a marathon. But the circumstances of her era, of that particular race, and of her incredible poise transformed her into an icon.

Myths exist to teach us lessons. David fought Goliath and etched into humanity's collective unconscious the idea that underdogs always stand a chance. The story of Icarus reminds us not to let our ego get in the way of our common sense. Adam ate an apple and introduced us to the perils of temptation.

Switzer's determination to finish her race even after being

attacked—and then continuing to fight for women to simply experience the joy of running for decades afterward—embodies fearlessness. And Switzer is just one of many mythological heroes. Here are a few more:

Jackie Robinson: Number 42 endured racist taunts and threats while becoming the first player to break the color barrier in Major League Baseball when he took the field for the Brooklyn Dodgers in 1947. Legendary Dodgers general manager Branch Rickey signed Robinson—but before he did, he spent hours trying to understand how Robinson would act when he encountered racist threats while on the field. "They'll taunt you and goad you," Rickey said during one meeting with Jackie.

At one point, Robinson interrupted the executive. "Mr. Rickey," he said, "are you looking for a Negro who is afraid to fight back?"

"Robinson, I'm looking for a ballplayer with guts enough not to fight back," Rickey replied.

"Mr. Rickey," said Robinson, "if you want to take this gamble, I will promise you there will be no incident."

Was Jackie Robinson passive or compliant? On the contrary, he quietly defied the social conventions of his time—Robinson never got in any incidents and let his play do the talking. Number 42 was himself an agent for change. After Robinson broke the color barrier in 1947, four more black players joined MLB rosters that year. Baseball—and sports—would never be the same.

The Lesson: *Bravery.* It might be terrifying to be a trailblazer, and it might attract hatred and other ugliness, but it's also the key catalyst for progress. Always let your play speak for itself, and keep going.

Serena Williams: Growing up, Serena Williams was smaller than everyone else. She was known as Venus Williams's kid sister. While filming a project for Nike, she once told me, "I always felt like I was just fighting to make it."

And so Serena trained. She trained and trained. "Training every day for thirty years is difficult," she once said. "Then I also think, *Someone is out there working really hard, and there's a poster of me, and they're working to beat me.*" Serena practiced every single day for thirty years. She's still doing it today, and I wouldn't be surprised if she was still doing it a decade from now. Nobody says she's too small or just a little sister anymore. They say she's the best woman to ever step on a court.

The Lesson: *Hard work.* The GOATs train and focus every day. By doing whatever it is that you love without compromise, you can become world class. The best part? That journey can start at any time. Why not today?

Steve Gleason: In 2006, during the New Orleans Saints' first game back in the Superdome following Hurricane Katrina, a practice squad player became one of the most important figures in Saints history. Steve Gleason, a backup to the backup, came barreling through the line and blocked a punt that rocked the Superdome and jolted the city back to life. Gleason quickly became a folk hero in New Orleans, immortalized by a nine-foot-tall bronze statue outside the stadium. But in subsequent years, Gleason became a different kind of hero. Paralyzed, with a diagnosis of amyotrophic lateral sclerosis (ALS), Gleason was given five years to live—max. To date, it's been almost a decade now since that prognosis, and Gleason is not only still alive, he's remained a powerful force in his community. His nonprofit foundation has raised so much money and awareness for ALS patients that in 2020, he was awarded a Congressional Gold Medal for his leadership.

I could try to describe the power of Gleason's world view. But why not let him tell you himself? This is from an open letter he wrote to the city of New Orleans:

> In 2011 I was diagnosed with a hurricane of a disease. ALS . . . Terminal. Death. 2–5 years. Like this city's levees in 2005, my invincible body has failed me, but like the residents of a city built 5 feet below sea level, I choose to be an idealist. We simply must be steadfast, maniacal idealists. When the world sees tragedy, idealists see opportunity. When the world folds its hand, idealists double down. When the world retreats, idealists reinvent. Idealism isn't for the fainthearted or weak-minded. ALS and the waters surrounding New Orleans have shattered our hearts a thousand times over. Somehow, like the local banana tree, our enduring hearts piece themselves back together each and every time. Rebuild. Rebirth. Repeat.

The Lesson: *Optimism.* Gleason has overcome tragedy in his life by refusing to accept defeat. It may seem trite, but the best athletes in the world never give up. When confronted with a challenge, they double down. When they feel a negative attitude sneaking in, they barricade themselves from it. As Gleason says, he's a "steadfast, maniacal idealist."

Chapter 4: Transcendence

What we learned: Ask nearly any great athlete about their most memorable moment on the court, field, pitch, etc., and you'll start to hear a lot of the same qualities: the action seemed to move in slow motion, their bodies animated on their own accord, everything felt effortless. They're all talking about being in the zone.

Nearly all religions try to help their followers experience this same state of consciousness, also known as "flow state." But unlike almost any other experience (music an analogous exception), in sports, flow state can be transferred from an athlete to a stadium full of fans to even a viewer watching at home. Simply *watching* a legendary performance can cause people to start to feel flow.

Here are some tips on how to optimize your opportunity to slip into the zone.

- *Focus:* Whether you're watching a game or playing a sport, be present. Ground yourself fully in the moment in order to experience the zone. If you're watching a game, stop checking your texts every five seconds, scrolling social media, or checking on your fantasy team. If you yourself are playing (or running, swimming, etc.), focus on your breaths. If your mind begins to drift into the past (regrets, emotions, etc.) or the future (anticipation, nervousness, etc.), gently bring yourself and your awareness to the present.

- *Find a challenge:* Even though all of the athletes I've spoken with have wildly different experiences for their greatest moment, none of them occurred totally randomly. Challenges that push us to our limits often help push us into the zone. So, watch your team in a big, important game. Or play against extra-talented players during pickup basketball, run an extra mile, or try a golf course where you can barely imagine yourself hitting par. It's when we come up against our physical and mental boundaries—or push beyond them—that we often discover our true potential.

- *Compete:* Competition is a primordial human instinct. Put two people at a starting line, let them gaze at the finish line, and yell "go." It's instinctive for them to try to beat their rival and bask in the glory of it. But the best athletes on the planet, those who have probably achieved fame and fortune through their success, discover that those rewards are fleeting. A higher purpose and adversary then comes into focus: themselves. Ask Tom Brady, Kobe Bryant, Serena Williams, Kelly Slater, or any other GOAT who their greatest rival is, and they will categorically come up with the same answer: they're in competition with themselves.

  Be in competition with yourself. Set a goal . . . and then seek to not just achieve it, but beat it. And when you do, reset and do it again. You'll be surprised by what you are capable of.

Chapter 5: Moral Codes

What we learned: In America each year, more than 42 million children participate in organized sports. While too many parents who put their children into youth sports programs do so with ridiculous dreams that one day their child will become the next Tiger Woods or Stephen Curry (they won't), most hopefully just want their kids to have fun—and learn some valuable life lessons along the way. This is at the heart of the Religion of Sports, the morality and ethics that encode the faith.

Teamwork, resilience, accountability, courage—these are but a few of the most valuable lessons that are omnipresent in sports. I told the story of my son, Krishu, who learned from his taekwondo instructor, Master Quan, the value of getting up off of the mat after being knocked down. As I reflect on my son's life to date, I realize that moment is as valuable as anything he's ever learned in or out of a classroom. For that reason, coaches often serve as a spiritual guide for how to best live a noble life. They are invaluable teachers, showing us the ways in which to navigate the challenges that come from competition, winning, and most importantly losing.

Below you'll find a list of some inspiring quotes from some of the greatest coaches of all time. See if you can guess which legendary sideline figure said each one. The answers are on the next page.

Quotes

1. "Winning is not a sometime thing; it's an all-time thing. You don't win once in a while, you don't do things right once in a while, you do them right all the time. Winning is habit. Unfortunately, so is losing."
2. "When a player makes a mistake you always want to put them back in quickly. You just don't berate them and sit them down with no chance for redemption."
3. "When you're in a situation, you can complain about it, you can feel sorry for yourself, you can do a lot of things. But how are you gonna make the situation better?"
4. "You can't force your will on people. If you want them to act differently, you need to inspire them to change themselves."
5. "If you start worrying about the people in the stands, before too long you're up in the stands with them."

Answers

1. Vince Lombardi
2. Pat Summitt
3. Tony Dungy
4. Phil Jackson
5. Tommy Lasorda

Chapter 6: Pilgrimage

What we learned: Both athletes and fans make pilgrimages to feel more connected to the Religion of Sports. Athletes work a lifetime to reach a championship game, while fans travel across the world to cheer on their favorite teams and witness sports' most storied traditions and cathedrals.

Sports can take you all around the world; they've certainly taken me places I would have never imagined. Take a look at all the places that sports can take you:

The Masters
The Indianapolis 500
The Daytona 500
Wimbledon
The Grand Sumo Championship in Japan
The Kentucky Derby
24 Hours of Le Mans
A Cricket Match at Eden Gardens
WWE Wrestlemania
A Red Sox Game at Fenway Park
A Cubs Game in the Bleachers at Wrigley Field

A Knicks Game at Madison Square Garden

The Olympics

An FC Barcelona Game at Camp Nou

A Real Madrid Game at the Bernabéu

A Bayern Munich Game at Allianz Arena

A Manchester United Game at Old Trafford

A Liverpool Game at Anfield

A Round of Golf at St. Andrews

A Round of Golf at Pebble Beach

A Montreal Canadiens Game at the Bell Centre

A Title Fight in Las Vegas

The Tour de France

The Monaco Grand Prix

Walk in the Footsteps of Legendary Athletes in Rome at
    the Circus Maximus

An LSU Game at Tiger Stadium

A Rose Bowl

The Baseball Hall of Fame in Cooperstown, New York

Sit with the Cameron Crazies for a Duke-UNC Game

Go to Omaha for the College World Series

A Rugby Match at Ellis Park Stadium

A Boca Juniors vs. River Plate Game in Buenos Aires

A Brazil National Team Game at the Maracanã

An El Tri Game at Estadio Azteca

A New Zealand All-Blacks Game at Eden Park

Skiing in Whistler, British Columbia

Surfing Pipeline on Oahu

The Running of the Bulls

Climbing Mount Everest

Chapter 7: Reformation

What we learned: Religions don't stay the same over time. Faith is contextual and adapts to the circumstances and challenges around us, providing insight and shining a light on the ways in which we continue to walk the righteous path. Ideally every generation of believers learns the tenets and traditions of their faith and reshapes it in new ways that suit their own times and experiences. Of course, change makes people uncomfortable and often invites debate for what a faith should be. It's through this debate, though, that a faith is tested and ultimately morphs into something even more relevant and existential. When Simone Biles pulled out of the 2020 Tokyo Olympics and admitted to her own mental health struggles, she ignited a cultural debate among those that praised her for candor and those that criticized her for a "lack of mental toughness." In the process, she became a leader of a reformation of the Religion of Sports.

For decades, locker rooms had been the domain of axioms like "Suck it up" and "Be tough!" Simone—along with athletes like Naomi Osaka, Kevin Love, and Mikaela Shiffrin—proved that it doesn't have to be that way. They showed that athletes can achieve great things and still prioritize their mental health.

Below you'll find a list of some of the most famous advice in sports history. Pick one that you think should remain influential, no matter what. Then pick one that we'd be better off without.

"Winning isn't everything, it's the only thing!"
"You miss one hundred percent of the shots you don't take."
"It's not about if you get knocked down. It's how you
get up that matters."

"Run when you can, walk if you have to, crawl if you
must; just never give up."
"Hard work beats talent when talent doesn't work hard."
"Don't run away from challenges. Run through them."
"You don't have to go fast, you just have to go."
"It is not the size of a man but the size of his heart that
matters."
"Never say never."
"It's not the will to win that matters—everyone has that.
It's the will to prepare to win that matters."

Chapter 8: Tribes

What we learned: Almost everything in America is designed to pull
people apart these days—especially if those two people are somebody
like me, an Indian-American from New England, and someone like
Johnny "Joey" Jones, a self-described "redneck" from Georgia. But
through our shared love of sports—and we aren't even fans of the
same sports—we have come to realize that we speak a similar language. We're still good friends.

In this chapter, I wrote that it's not that there is a Religion of the
Red Sox or a Religion of the Yankees. These are just two of the many
tribes (or denominations) that make up the larger Religion of Sports.
I want you to think about your absolute favorite team in the world.
Now think about the elements that make your tribe unique among
every other franchise in sports. In sum, you will prove to yourself that
your fandom is a faith.

Use the chart that follows to help find the elements that make
your tribe unique.

Team name: _____

PROPHETS

Who are the players, coaches, executives, fans, announcers, or other characters that help define your tribe? Put another way: Whom would you put on your team's Mount Rushmore? This isn't necessarily a list of the best figures ever, but rather the ones that most define what it means to be a member of your tribe. (E.g., for the Chicago Cubs, the list might include: Ernie "Mr. Cub" Banks, third baseman Ron Santo, announcer Harry Caray, and songwriter Steve Goodman.)

_____

_____

_____

_____

MIRACLES

What's the most magical moment in your team's history? (E.g., a Los Angeles Dodgers fan might point to Kirk Gibson's home run in the 1988 World Series.)

_____

_____

_____

RELICS

What's an item that has come to have outsize and holy importance for your tribe? (E.g., the University of Kansas Jayhawks have a copy of James Naismith's original rules of basketball in a museum on campus.)

_____

HYMNS

What song does every member of your tribe sing together? (E.g., Liverpool has adopted the Rodgers and Hammerstein show tune "You'll Never Walk Alone" as the Reds' hymn.)

_____

SYMBOLS AND GESTURES

What are the little cues that somebody is a member of your tribe? (E.g., Green Bay Packers fans wear cheeseheads at Lambeau Field to show their allegiance to the Pack.)

_____

_____

_____

RITUALS

What's a tradition that's practiced by your tribe and nobody else? (E.g., Detroit Red Wings fans throw squid on the ice whenever a player records a hat trick.)

_____

_____

CATHEDRALS

Which places do your tribe consider to be sacred ground? It might not be just your stadium—maybe there's a special bar or restaurant, too. (E.g., less than a mile from the University of Texas Longhorns' football stadium sits Scholz Garten, the oldest operating business in Texas. In 1893, the first unbeaten Longhorns football team celebrated their season there—and to this day, it is filled with burnt orange on game days.)

_____

_____

HOLY WAR

What's a team that your tribe just can't stand to see win? (E.g., Duke vs. North Carolina.)

_____

Chapter 9: Afterlife

What we learned: Kobe took parts of his game from all the basketball greats who came before him. He studied Michael Jordan and mastered footwork. He looked at Oscar Robertson for inspiration on how to use his body in the paint. He obsessed over the way that Larry Bird used his eyes during his pump fake.

Now players in today's NBA grew up taking their inspiration from the Black Mamba. I'd like you to go on a little scavenger hunt—all in the name of religion. Watch an NBA game (if it's not during the season, find some highlights from last year) and see if you can find three times where a player makes a move inspired by Kobe. Here's a hint: in 2020, ten NBA players wore numbers in honor of Kobe.

Not a basketball fan? No problem. Pick your favorite sport and think about one of its gods: Sandy Koufax in baseball, Diego Maradona in soccer, Walter Payton in football, Muhammad Ali in boxing, or Martina Navratilova in tennis. Next time you watch a game or a match, search for all the ways that they've influenced their sport to be what it is today.

If you do that, you'll see how sports can help its legends live on forever and shape the religion in perpetuity.

But now comes the really fun part. Go watch a game. Go kick a ball around. Go cheer on your favorite team in the stands. The Religion of Sports, like every other faith, is ultimately its believers—and now, I hope that you'll count yourself as someone who views sports as nothing short of a spiritual exercise as well.

See you in the stands. See you in church.

Acknowledgments

I am not quite sure where to even start with a list of acknowledgments for this book. I have so much gratitude in my heart for the book you have in your hands, and the moment I take that from the conceptual to the literal, I fear that it will be contained and forsake many who have played a part. . . .

Alas, let's start here: my cowriter for this book, Joe Levin, was in eighth grade when I think I first started really conceiving the conceit for the Religion of Sports. He was a junior high kid overhearing me and his dad, Jordan, kicking around this idea. Over the next ten years, as the idea morphed into a show, then a production company, a content and media studio, and now into a book, Joe has been a voracious and hungry protégé and now partner. Without him, this book simply would not have been possible. Thanks, Joe.

Michael Strahan and Tom Brady. These guys also came aboard the Religion of Sports train when it was just an idea. It wasn't because I showed them a business plan or a ten-year road map that would take us here. It was because they believed in it, and they believed in me.

Among the many athletes whom I have worked with and who really helped shape the nucleus of the Religion of Sports, I need to call out one: Kobe Bryant. Every day with Kob was a fistfight, a combative collision of egos, ideas, and creativity. *I miss you, my friend, and I thank you.*

From there, it's really impossible to articulate all the folks who have been a part of this journey. Literally hundreds of creators, producers, editors, writers, cinematographers, agents, subjects, and others who have been a part of the alchemy that has forged the thesis that sports is a religion. I can't call out every single person who has contributed, but I will call out a few:

Ameeth Sankaran, my business partner and friend, who has taken my endless ramblings, organized and codified them, and enabled us to engage so many incredible people to reimagine the Religion of Sports over the years. Jay Adya, Nikhil Bahel, Shari Redstone, Ashish Patel, Andy Howard, Sumant Mandal, Varun Soni, Gene Wolfson, Pete Mattoon—you all have been true believers. That's the fabric of faith.

And now a few of my collaborators through the years: Martin and Matt Roe, Chris Uetwiller, and their wildly talented team at Dirty Robber. The act of nothing becoming something is magic, and I am eternally grateful.

They call it "unscripted programming" for a reason. Because no matter the plan, it always goes awry. So who is gonna be standing shoulder to shoulder with you to figure it out? To problem-solve and proceed? For me, it's these folks: Victor Buhler, Giselle Parets, Erik LeDrew, Meg Cirillo, Sean Horvath, Alex Trudeau Viriato, Brady Hammes, Kevin Barth, Logan Fonseca, Lauren Fisher, Chelsea Marotta, Adam Schlossman, Michael Bloom, Justin Skelly, Pam Daniels, Eric Hoberman, Dan DiStefano, Jacob Mosler, Radhika Womack, Chris Madrigal, Lilly Flaherty, Jessica Young, and many, many more.

Okay, so you have great ideas. And you have great partners to help you make them. How are you going to get them out? It's the people who stick by you when the world says, "Nah. Next." That's how. For me, that's been Josh Pyatt and Jason Hodes at William Morris Endeavor. Aaron Marion, who's the big brother ensuring everyone knows about everything once it exists. Amanda Annis, who persisted the three times this book got canceled and pushed it through anyway. Amar Deol, our editor, who picked us back up and brought shape to what you just read. Ben Rawitz and Constance Schwartz Morini—the friends and colluders behind the scenes who never demand credit, but make everything go and deserve all of it.

And then there's the family. Papa, who exposed me to religion before there was the Religion of Sports. Mom, who without even knowing it was the first one who instilled in me the faith, not because she loved sports, but because she loved me. My sister, Mallika, who indulged my obsessions with all things Red Sox, Bruins, Patriots, and Celtics growing up, while she couldn't have cared less for them. The next generation—Tara, Leela, Kiran, and Alex. Noah—that you have become a Bruins diehard is the most unexpected and greatest gift you could ever give me. Krishu—that you ride and die with me and the Celtics despite having been born and bred in Santa Monica and being a Southern Cali kid is absolute affirmation of everything to me. I love you more than I love the Red Sox's winning their first chip in 2004, which says it all.

Candice—my muse. In 2003, when Aaron F'ing Boone and the damn Yankees broke my heart again, you were the only reason it kept on beating. You told me after that game that you couldn't understand why I cared so much for this thing I had no control over. I told you then that I'd figure out a way to explain it to you. And the seed for the Religion of Sports was planted. This is, and I am, because you are. I adore you.

About the Author

Gotham Chopra is an Emmy Award–winning filmmaker and the cofounder of Religion of Sports. He directed *Man in the Arena: Tom Brady*, *Kobe Bryant's Muse*, *Simone vs Herself*, and countless other documentaries, having worked with many of the greatest athletes of all time. He is also the author of *Walking Wisdom* and *The Seven Spiritual Laws of Superheroes*. Gotham lives with his wife and son in Los Angeles, where he still worships his beloved Boston Celtics even among the philistine Laker fans.

Joe Levin is a writer who lives in Austin, Texas, via the bleachers of Wrigley Field. His writing has appeared in *Texas Monthly*, Wildsam, and more.